Internal Displacement and Conflict

Grounded in multidisciplinary research, this book presents a methodical understanding of those displaced within their national borders, the Internally Displaced Persons (IDPs).

The IDP phenomenon remains less understood than that of refugees due to the "internal" nature of the crisis, linked to a nation's sovereignty, which assigns the responsibility for care to the national actors as opposed to an international body. However, the IDP phenomenon poses an international humanitarian challenge, with upwards of 40 million people currently in internal displacement across the globe. This book helps answer the most perplexing questions surrounding conflict-induced protracted displacements: namely, how do positions embraced by key actors inform/influence IDP policies, and why, despite the promise of robust return packages, do families remain reluctant to return to home communities and equally reluctant to embrace new host communities? Capitalizing on the diagnostic tool kit known as Dugan's Nested Model, uniquely adapted to the Kashmiri Pandit displacement, this book also analyzes issues of the similarly displaced communities of Nagorno-Karabakh, Abkhazia and South Ossetia, Kosovo, and Darfur regions.

This book will be of much interest to students of peace and conflict studies, humanitarianism, Asian politics, and International Law in general.

Sudha G. Rajput is Adjunct Professor at the School for Conflict Analysis and Resolution, George Mason University, Virginia, USA.

Internal Displacement and Conflict
The Kashmiri Pandits in Comparative Perspective

Sudha G. Rajput

LONDON AND NEW YORK

First published in paperback 2024

First published 2019
by Routledge
4 Park Square, Milton Park, Abingdon, Oxon OX14 4RN

and by Routledge
605 Third Avenue, New York, NY 10158

Routledge is an imprint of the Taylor & Francis Group, an informa business

First issued in hardback 2019

Copyright © 2019, 2024 Sudha G. Rajput

The right of Sudha G. Rajput to be identified as authors of this work has been asserted by her in accordance with sections 77 and 78 of the Copyright, Designs and Patents Act 1988.

All rights reserved. No part of this book may be reprinted or reproduced or utilised in any form or by any electronic, mechanical, or other means, now known or hereafter invented, including photocopying and recording, or in any information storage or retrieval system, without permission in writing from the publishers.

Trademark notice: Product or corporate names may be trademarks or registered trademarks, and are used only for identification and explanation without intent to infringe.

Publisher's Note
The publisher has gone to great lengths to ensure the quality of this reprint but points out that some imperfections in the original copies may be apparent.

British Library Cataloguing-in-Publication Data
A catalogue record for this book is available from the British Library

Library of Congress Cataloging-in-Publication Data
Names: Rajput, Sudha G., author.
Title: Internal Displacement and Conflict: The Kashmiri Pandits in Comparative Perspective / Sudha G. Rajput.
Description: First edition. | London; New York, NY: Routledge/Taylor & Francis Group, 2019. |
Includes bibliographical references and index.
Identifiers: LCCN 2018054389 (print) | LCCN 2018056618 (ebook) | ISBN 9780429764639 (Web PDF) | ISBN 9780429764622 (ePub) | ISBN 9780429764615 (Mobi) | ISBN 9781138354265 (hardback) | ISBN 9780429427657 (e-book)
Subjects: LCSH: Internally displaced persons–India–Social conditions. | Kashmiri Pandits–Social conditions. | Internally displaced persons–Social conditions.
Classification: LCC HV640.5.I4 (ebook) | LCC HV640.5.I4 R35 2019 (print) | DDC 362.87088/294509546–dc23
LC record available at https://lccn.loc.gov/2018054389

ISBN: 978-1-138-35426-5 (hbk)
ISBN: 978-1-03-292956-9 (pbk)
ISBN: 978-0-429-42765-7 (ebk)

DOI: 10.4324/9780429427657

Typeset in Times New Roman
by codeMantra

I dedicate this book to my husband, Mr. Rajinder S. Rajput, whose unwavering support throughout my doctoral studies has made it possible for me to contribute this important piece of literature to bring a systematic and scholarly understanding of conflict-induced internal displacements.

I also dedicate this monograph to the numerous Delhi- and Jammu-based displaced Kashmiri Pandit families who welcomed me into their camps, homes, and shops to make it possible for me to collect vital research data. Their personal stories have touched my heart and form the backbone of this study.

Sudha G. Rajput

Contents

Foreword ix
Acknowledgments xi
List of abbreviations xii

SECTION I
Global phenomenon of internal displacement,
Kashmiri Pandits, research challenges,
and family legacies 1

1 Global phenomenon of internal displacement
  and Kashmiri Pandit community 3

2 Context: internal displacement of Kashmiri Pandits 21

3 Challenge of researching protracted displacements 26

4 Oral account of one Kashmiri Pandit family
  and the legacies left behind 34

SECTION II
Kashmiri Pandit challenges and the dilemma of return 43

5 Methodical analysis of Kashmiri Pandit challenges 45

6 Moral and political dilemma of return 57

SECTION III
Policies, assessment, positions, and complexity of policymaking   67

  7 Kashmiri Pandit families evaluate "migrant" policies   69

  8 How actor positions influence policy outcome   76

  9 Complexity of IDP policymaking   82

SECTION IV
Understanding Kashmiri Pandit displacement through a comparative lens: journeying into Azerbaijan, Georgia, Serbia, and Sudan   89

10 Azerbaijan: displaced from Nagorno-Karabakh   91

11 Georgia: displaced from Abkhazia and South Ossetia   101

12 Serbia: displaced from Kosovo (ethnic Serbs)   112

13 Sudan: displaced from Darfur   123

SECTION V
Findings, best practices, and moving forward   137

14 Findings, best practices, and moving forward   139

    Annex 1: Fieldwork log   153
    Annex 2: Kashmiri Pandit Families Assess
       'Migrant' Policies   158
    Index   161

Foreword

Sudha G. Rajput has written a book that gives us great insight into the plight of Internally Displaced Persons (IDPs) in five cases. Focused particularly on the displaced Hindu Pandits (KPs) of Kashmir, forced out by rising anti-Pandit violence in the Valley in 1989, the analysis includes case studies of IDPs in Azerbaijan, Georgia, Serbia, and Sudan (Darfur). These are among the 40 million people estimated to be internally displaced within their own countries of origin. They are distinct in several ways from the more widely discussed groups categorized as "refugees," those displaced outside the boundaries of their nation-state. The latter enjoy some measure of protection or representation by international organizations such as UNHCR. IDPs, presumed to be under the care and protection of their state, are in many senses on their own. Rajput is skillful in showing how this is both a bane and a blessing. Indeed, one of the strengths of her analysis is in demonstrating (particularly in the KP case) how communities of IDPs, long separated from their homes, with doubtful chances of return, seize as much control over their lives as they can, demonstrating agency and strategy in dealing both with host communities (co-citizens and often co-ethnics, but with complicated relations nonetheless) and governmental (state) agencies.

The cases that Rajput investigates are those where displacement is the result of internal, usually violent, conflict, not, for example, displacement due to natural disaster (earthquake, flooding) or intentional, state-initiated grand projects (hydroelectric dams). Such conflicts include civil wars, secessionist movements, and large-scale anti-state insurgencies. In a way, these causes have already challenged the sovereignty of the state, its endurance or its legitimacy, and the IDPs can be construed as the human embodiment of that challenge. In Rajput's words, they present by their existence a "moral dilemma and hazard" for the state. The state not only has reasons to protect and support its IDP citizens but also has reasons to "mask" the roots of their plight, labeling them "voluntary migrants" or victims of a "temporary disturbance."

Rajput demonstrates convincingly that how the (state) "elites" or policymakers initially label or categorize the IDPs will determine much in the way of how they are subsequently treated. At the same time, as IDPs realize their internal "exile" is likely to be a long (or even permanent) one, some of them are able skillfully to use the rhetorical dissimulation of the state's narrative to extract benefits. Of course, they too are victims of this narrative dilemma: a longing for their homeland, together with the trauma of recollecting the humiliation of their exile, and the feeling, especially among the young, that forced exile may in fact have been a great "blessing in disguise."

Rajput moves not only across cases but also across societal levels within cases: comparing national elite policy and institutional actors with the lived experience of individual IDPs and families "on the ground." She ranges from ethnographic participant-observation and oral history to policy analysis and best practice recommendations. She rightfully includes how the IDPs articulate with their host communities as one of the crucial determinants of the quality of their everyday experience and their future. The result is an important work that enhances our understanding of conflict-induced displacement in comparative, institutional, and human perspectives.

Kevin Avruch
Dean
Henry Hart Rice
Professor of Conflict Resolution
Professor of Anthropology
School for Conflict Analysis and Resolution
George Mason University

Acknowledgments

First and foremost, I acknowledge the guidance received from the academic community of the School for Conflict Analysis and Resolution, George Mason University. The candid reviews received from its many scholars are largely responsible for positioning the displacement of the Kashmiri Pandit community in a comparative perspective, thus bringing a global focus to the issue and to the book.

I am indebted to Dr. Kamal Pandit's family of the Jammu Jagti Camp. Being embedded within his family provided immense insights into the hidden aspects of the community's displacement, facilitating my ethnographic approach. I hold dear the love and affection of the Pandit family. I am also grateful to the numerous other displaced families in Jammu, Srinagar, and Delhi that welcomed me into their homes, shops, and schools, and helped me learn firsthand of the events described in the book.

I owe gratitude to Dr. Sazawal, President of Kashmir Overseas Association, whose exclusive insights on Kashmiri Pandits helped draw my attention to the existing scholarship and to the role of civil society.

I am grateful for the many Delhi- and Kashmir-based policymakers, advocates, nongovernmental organizations, and journalists who welcomed me into their offices to help bring a policy perspective to displacement; a large part of this communication network was facilitated through the support of Ms. Ashima Kaul, a Delhi-based journalist.

I am forever grateful to my husband, Mr. Rajinder S. Rajput. His unwavering support and his complete faith in my capabilities have resulted in this much-needed publication.

I acknowledge the continued support of my children and their spouses: Sura, Amit, Sandeep, and Supriya. They were often the testing ground for my formative ideas. I am equally grateful to my Delhi-based family, who helped shine a local perspective on KP displacement, and to Mr. Gul Chauhan, who accompanied me to Jammu and Srinagar.

Sudha G. Rajput
Washington, D.C.
September 2018

Abbreviations

AIKS	All India Kashmiri Samaj
DDA	Delhi Development Authority
EU	European Union
FMR	Forced Migration Review
FYR	Former Yugoslav Republic
ICRC	International Committee of the Red Cross
IDMC	Internal Displacement Monitoring Center
IDPs	Internally Displaced Persons
INA	Indian National Army Market (location of Kashmiri "Migrant" shops in Delhi)
J&K	Jammu and Kashmir State (India)
KP	Kashmiri Pandits
NATO	North Atlantic Treaty Organization
NGOs	Non-Governmental Organizations
OSCE	Organization for Security and Cooperation in Europe
OCHA	Office for the Coordination of Humanitarian Affairs
SCR	Serbian Commissariat for Refugees
SFRY	Socialist Federal Republic of Yugoslavia
UNAMID	United Nations African Mission in Darfur
UNHCR	United Nations High Commission for Refugees
USAID	United States Agency for International Development
Valley	Refers to Kashmir Valley

Section I

Global phenomenon of internal displacement, Kashmiri Pandits, research challenges, and family legacies

1 Global phenomenon of internal displacement and Kashmiri Pandit community

Grounded in multidisciplinary and multipronged research, this book presents a methodical and nuanced understanding of those who remain displaced within their national borders, known within the international context as the *Internally Displaced Persons* (IDPs). This sharply contrasts with the wide coverage of other displaced groups, such as the refugees or the asylum seekers, who are covered under a well-structured system of international care and protection. The IDP phenomenon remains less understood, given the internal nature of the crisis, linked to a nation's sovereignty, which assigns the responsibility for care and protection to national actors instead of an international body, such as the United Nations High Commission for Refugees (UNHCR), that oversees the refugee population. In the interest of keeping the displaced populations protected from further harm and the larger populations uninterrupted, often, the national actors feel compelled to downplay the IDP crisis by either reframing it or disguising it in ways that keep the displaced populations hidden and the larger populations uninformed, and thus unconcerned about those who become internally displaced.

However, the magnitude of the IDP phenomenon poses a universal humanitarian challenge, with upwards of 40 million people currently in internal displacement, enduring an average length of 17 years of remaining outside of their home communities (IDMC, 2016, OCHA, 2015). Astonishingly, as large as these numbers are, they do not account for the fact that many of those displaced go into hiding due to the stigma of displacement, thus escaping the registration process. The *Guiding Principles on Internal Displacement* clarify the concept of IDPs as applying to

> persons or groups that are forced or obliged to leave their homes and who seek shelter within their country, as opposed to the

refugees who may be forced or obliged to leave their homes for the same reasons but who cross an internationally recognized border in order to secure shelter.

(UNHCR, 2004)

The labeling of those who flee or are forced to flee into these two distinct categories has deeper implications, which come to impact not only people's rights to return and rehabilitation but also their standing in the *host* communities, thus impacting their everyday lives. The host communities are the new societies where the families come to seek shelter or are directed to do so by the responsible agencies.

Among the forces that result in internal displacement are ethnic violence, clashes between the majority and the minority groups, civil wars, political unrest, the actions of oppressive regimes, the rise of militancy, illegal seizure of land, natural disasters, and a nation's own development projects and agenda.

This book addresses what appears to be the most troubling of these forces: namely, the conflict-induced displacement, which accounts for an estimated 38 million of all those living in internal displacement (BBC, 2015). Often rooted in forced evictions by members of one's own society, such displacement leaves enduring scars of shame, humiliation, and loss of moorings, not only affecting those displaced but also spilling onto a whole new generation born and raised inside the camps, settlements, and compartmentalized communities. Communities displaced by conflicts are never fully accepted either, in host communities or on return to their own ancestral homes. Paradoxically, these groups remain tied to the forces that displaced them in the first place. Entrapped in political and community parameters of government-designated housing or remaining in the shadows of their host communities, these families endure an upsetting process while remaining in internal exile indefinitely. Given their prolonged physical absence from home communities and the consequent loss of many of the rights afforded to them as residents of those communities, these families live with minimal hope of return to hometowns and at best a lukewarm acceptance by their host communities.

There have been many such displacements across the globe, such as the displacement of the 250,000 Kashmiri Pandit (KP) families (ACCORD, 2010), a high-caste Hindu society ousted from their ancestral homes in the Kashmir Valley in 1989 (this displacement serves as the primary case for this book); the displacement of the families from the Nagorno-Karabakh area of Azerbaijan, where close to half a million people remain displaced after more than two decades (IDMC, 2014); the displaced from the Abkhazia and the South Ossetia regions of Georgia, ousted from the areas which became the targets of secessionist influence in the early 1990s; the ethnic Serbs displaced from Kosovo and now placed under the rubric of Serbian IDPs; as well as those displaced from Darfur in Sudan, where the violence that erupted in 2003 had displaced more than three million people (Darfur genocide, n.d.).

Key questions addressed

This book helps answer two of the most enduring and perplexing questions surrounding conflict-induced protracted displacements:

- How do the *positions* embraced by key displacement actors inform and influence IDP policies?
- Why, despite the promise of robust return and rehabilitation policies and packages, do the families remain reluctant to return to home communities and equally reluctant to embrace their new host communities?

Principal argument

The probe into the elite positions unveils that regardless of the duration of displacement, the officials steadfastly hold on to the initial storyline that they use in order to explain the initial crisis to the wider population as it unfolds. Consequently, any reforms made to the IDP policies throughout the phases of displacement are intended to align with and reinforce these very initial positions. Therefore, it is concluded that the IDP policy formulation remains a direct function of how the officials address and *position* (Harre & van Langenhoven, 1999) the crisis as it develops.

Theoretically, the initial stance used to address, interpret, and manage the crisis is meant to provide instant and short-term relief to the families as they arrive in new communities. However, in practice, as displacement protracts, the officials stay firmly grounded in their initial explanations, fearing the public outcry due to any shifting narrative.

The initial positions also serve to protect the elite from formulating long-term policies, thus resulting in a portfolio of ongoing ad hoc temporal policies, often ill-suited to address the needs of those who remain in long-term displacement.

This book exposes that the narratives formulated to explain the initial crisis to the public, such as by identifying it as an outcome of "a temporary disturbance" or "a voluntary migration" (personal communication, August 1, 2011, Delhi-based official) and the labeling of the families accordingly, pose a moral dilemma and hazard not only for those displaced but also for the policymakers. Furthermore, the same elite positions transfer to the psyche of the members of the host communities, who capitalize on these narratives and set in motion their own "power relations, identifying rules for inclusion and exclusion" (Tilly, 2005), thus governing how the displaced are to be treated in their communities.

The insights gained through this investigation not only offer a unique lens to probe the mind-set of the policymakers, and thus understand the hidden obstacles to the challenging task of IDP policymaking, but also offer valuable understanding of the politics as well as the intricacies of IDP/Host dynamics that come to govern the community's power relations. A systematic grasp of the schema of IDP challenges secured through this process offers possible approaches to the much-needed, context-driven societal and policy solutions.

Scope

As there are ample studies covering the issues of refugees around the world, including the Kashmiri refugees, the goal here is to shed exclusive light on the populations that often remain hidden and less understood. In an attempt to fill this painful void in internal displacement literature, the effort has carefully excluded the discussion of the Kashmiri or any other refugees. Therefore, the book demystifies the often-confusing concerns of the conflict-induced internal displacement of the KP community and compares and contrasts it with the similarly displaced communities of Azerbaijan, Georgia, Serbia, and Sudan's Darfur families. Consequently, the book unfolds the political, socioeconomic, and psychological dimensions of those who remain internally displaced while probing the mind-set behind policymaking; scrutinizing the intricacies of IDP/Host dynamics; and decoding the dilemma of return faced not only by those displaced but also by members of the host communities as well as the lawmakers themselves.

Key themes

The key themes explored in the book are:

Challenges of investigating the IDP communities

The challenges identified in the pages ahead provide clues into the less understood nature of the IDP phenomenon and help sensitize the readers to the hidden challenges of investigating these communities, exaggerated by the protracted nature of such crises. Specific investigative hurdles help unfold the political and the community positions embraced by key actors (Rajput, 2012) that often come to stifle scholarly and meaningful investigation of such sensitive communities.

Family challenges

A methodical analysis of the family challenges, employing the *Nested Model* (Dugan, 1996), helps illustrate the interconnectivity of the overlapping displacement issues, underscoring the need for responding to such crisis with a similar holistic mind-set.

Moral and political dilemma of return

The issue of return is illustrated to be the most troubling of the IDP challenges. In the event of violence-trigged displacement, the return to family homes becomes an issue of a political, legal, economic, and moral nature. The matter of return to home communities not only holds the families in *cognitive dissonance* (Festinger, 1957) that comes from a psychological tension of wanting to embrace their new society while fearing the lessening of loyalty and love for their home community but also keeps the policymakers on the defensive, leading to a portfolio of random and temporal policies, which come and disappear, while families remain displaced.

Policy assessment

The calculus, shaped by the families' own displacement-based, lived-in experience, provides an extraordinary understanding of how the displacement experience comes to be assessed as either a "blessing in disguise" or a "humiliation to the very being," illustrating the complexity of IDP policymaking and the mind-set behind the transitional and ad hoc policies. A considerate understanding of the bottlenecks to IDP policymaking not

8 *Global phenomenon of internal displacement*

only brings out nuances in an understanding of those who must endure displacement but also brings to bear an appreciation of the national policymaking process, hinting at a thoughtful and context-specific engagement of international actors in situations of internal displacement.

Primary case

As its main focus, this book takes up the displacement of a quarter of a million KPs of the Kashmir Valley (Valley), which ruptured the very fabric of this community in 1989. The KP displacement, treated under the rubric of "sensitive issues" and within the context of the wider "Kashmir problem," has kept researchers from scrutinizing either the triggers or the aftermath of this displacement. The dubbing of it as a "voluntary migration" (Rajput, 2012) has also kept the research community from conducting a meaningful examination into the displacement and exploring it as an issue of forced eviction, and thus of a humanitarian concern. Consequently generally documented to be the toughest misfortune of the overarching Kashmir issue, after nearly three decades, only a handful of issues of the KP displacement have become known, and those through the story-bound literature, while critical issues have remained hidden or misunderstood. The issues that have remained misunderstood pertain to the policymakers' intent to label this crisis as a "temporary disturbance"; the labeling of the KP community as "migrants"; or the puzzle of why, despite the attractive "return and rehabilitation packages," most families have not been able to return. Additionally, what remained hidden were the roles of KP-supported advocacy groups and Kashmir's civil society, as well as the roles and positions of the national leaders in the interpretation and the management of this crisis.

The book unfolds the political, socioeconomic, and psychological dimensions of this community, yielding a holistic understanding of the protracted displacement of the KP families. These clues provide a unique lens into the political and community entrapment of these families and into the politics and the intricacies of the IDP/Host dynamics.

Comparative cases

While recognizing and respecting that the IDP challenges are "unique to the culture and the nature of the hardships that the families encounter and endure" (Avruch, 2011), however, in order to situate the displacement of the KP community in the context of similarly displaced communities by drawing commonalities and individualities, issues of communities displaced by similar forces around the globe are also methodically analyzed. Therefore, capitalizing on the diagnostic tool kit uniquely developed (Rajput, 2012) to investigate the Kashmiri displacement, the book explores and analyzes issues of those displaced from the Nagorno-Karabakh area of Azerbaijan; those displaced from the Abkhazia and the South Ossetia regions of Georgia; the ethnic Serbs displaced from Kosovo; and the families displaced from the Darfur region of Sudan. These cases bear resemblance to the displacement of the KPs as the triggers of displacement in all these cases are conflict induced rather than natural disasters; the protracted nature and magnitude of displacements are also comparable, thus placing the level of human suffering of these communities on an equal footing. This comparative indulgence represents a unique effort in internal displacement literature toward a possible attempt to zero in on the core IDP experience, thus offering multiple context-fitting approaches into societal and policy solutions.

Scholarship

The account of the KP displacement is informed by a field-based triangulated methodology that collected data from all key actors of displacement: namely, the KP displaced families, members of the host communities, and the policymakers (Rajput, 2012). Ninety-four face-to-face interviews were conducted in Delhi, Jammu, and Srinagar, and included residents of 15 "Migrant" camps; shopkeepers of the "migrant markets"; government ministries; and members of civil society, media, academia, and medical and charitable organizations (Annex 1). The investigation involved interacting with 192 local residents. Two oral histories of the displaced families, constructed as a result of being embedded with the IDP families, inform how the families come to assess various policies as guided by their lived-in displacement experiences. The account of Georgia's displacement is informed by face-to-face interviews (Rajput, personal communication, 2015, Ministry of Internally Displaced Persons from the Occupied Territories, Accommodation and Refugees of Georgia) with Tbilisi-based policymakers and the displaced families residing at the Tserovani settlements on the outskirts of Tbilisi, Georgia. Knowledge of those displaced within Serbia is guided by field-based interviews (Rajput, personal communication, May 6, 2016, Serbian Commissariat for Refugees, Belgrade, Serbia) with Belgrade-based officials of *Serbia Commissariat for Refugees* and the displaced families, who fled from Kosovo in 1999 and now falling within the rubric of the Serbian IDPs living on the outskirts of Belgrade. The situation of those who fled from Darfur in 2005 and are now dispersed around Khartoum, Sudan, is informed by field-based interviews with the academic community in Khartoum (Rajput, personal communication, 2015, Academic community, University of Bahri, Khartoum, Sudan).

Theoretical approach

Given the protracted nature of the KP displacement, the individual issues of the families have come to be meshed and embedded into other issues, thus obscuring a clear understanding of the families' challenges. As the KP policy portfolio has remained woven around a singular "migrant" identity, the Nested Model (Dugan, 1996) has been adapted (Rajput, 2012) to systematically untangle the intertwining issues of the psychosocial, environmental, institutional, and other dimensions that have come to be enclosed within this singular identity. The Nested Model is used in social sciences to analyze conflicts as products of "events occurring at various interactive levels." In this framework, a conflict is understood as a comprehensive system, comprising of four circles, surrounded by progressively larger circles (Figure 5.1). Of immediate relevance is a *specific* issue, which itself is embedded in elements of a *relational* nature, the latter being entrenched within the larger context of *subsystem* issues, and the deepest and the underlying layer of the conflict system residing in the society-wide *structural* issues (Prendergast, 1997). This interconnectivity of issues creates a complex web, suggesting the importance of addressing the underlying elements of any given crisis. Engagement of the Nested Model has not only revealed unique features of the KP displacement but also successfully unfolded the overlapping issues of the IDP crisis in general.

Current debate on IDPs

The IDP phenomenon is not new, and statistical data on those displaced has been methodically collected and monitored at the global level, such as by the Internal Displacement Monitoring Centre (IDMC), since 1998. However, the magnitude of those who remain in protracted displacements and specifically conflict-induced displacements has been growing in recent years, pending the resolution of political and ethnic conflicts that have become more prevalent post the end of the Cold War. The *Guiding Principles on Internal Displacement* (UNHCR, 2004) have consistently maintained that the responsibility for the care and protection of those who remain within the national borders falls on the national actors. However, in the absence of universally enforced rules, the national actors are at liberty to formulate IDP policies as they see fit. As the return of the IDPs to hometown communities involves a tricky mix of approaches that entail reconciling with the perpetrators, rebuilding the war-torn communities, establishing law and order, and eradicating militancy, often, national governments remain reluctant to take on these controversial tasks and instead resort to keeping the families secure in host communities through temporary housing, jobs, and school quotas. In light of the growing protracted displacements, current debate on internal displacement is now focusing on establishing frameworks for durable solutions, where three settlement possibilities are being considered: the first and the preferred option calls for the safe return of the displaced families to hometown communities, the second option calls for the integration of the families into existing communities, and the third option requires moving the families into other communities within national borders.

Book's contributions

This book opens a dynamic platform for multilevel public debate where the context-sensitive societal and policy alternatives, proposed through the current debate, can be explored from the perspective of key actors of displacement. In the context of long-term displacement, such a platform is hoped to elevate the issue of IDP care and protection from a simple humanitarian gesture to one of self-reliance. For the first time, the adaptation of the Nested Model offers a unique diagnostic tool kit to unravel the often confusing and overlapping challenges of the displaced communities around the globe. The issue of displacement can now be understood as falling within the domain of "complex conflict issues," where displacement can be understood as both a consequence and a cause of further volatility in a society. The multipronged understanding of displacement now encompasses an account of the underlying positions and motivations of key actors, the roles of civil society and national and international actors, the intricacies of IDP/Host dynamics, and the meaning of return. The comparative aspect of this book offers a unique learning opportunity. As the book unfolds a systematic account of several communities, the global phenomenon of displacement can now be understood in terms of a comprehensive system, rooted in the structures of society and necessitating a life of lasting transitions of a complex nature. This contrasts with an understanding of displacement as an isolated issue, stemming from a stand-alone crisis of a social or a political nature and resulting in economic hardships. A comprehensive explanation suggests a similar holistic mind-set that allows for alternative approaches into the societal and policy solutions.

Prior scholarship that has contributed to the issue-specific understanding of KP displacement, such as Evans (2002), Oberoi (2004), Sekhawat (2009), and Moza (2012), is appropriately leveraged and rightfully credited throughout this book.

However, an absence of a scholarly and a holistic probe into this community has resulted in publications with random explanations of this community's predicament as well as the efforts of the officials. A nuanced understanding of the mind-set behind policy formulation and the intricacies of the IDP/Host relations brings to light not only the challenges of each actor but also an appreciation of the complexity of policymaking and the dilemmas faced by each constituency. This helps reiterate the need for all key actors, including international actors, to recognize gaps between national and international response, and to collectively explore possible approaches for durable and sensible solutions.

Book layout

This book is divided into five sections, with 14 chapters.

Section I: Global phenomenon of internal displacement, Kashmiri Pandits, research challenges, and family legacies

Section I quickly brings the readers up-to-date on the global phenomenon of internal displacement, and specifically the conflict-induced displacement of the KP community. Situating the text in its historical context facilitates a sound grasp of the rich text ahead. Next, the challenges of investigating internal displacement in general and the KP community in particular sensitize the readers to the hidden challenges of investigating sensitive and, what come to be, hidden communities. This enhances an appreciation of the depth and breadth of the research process required for conducting meaningful research in such delicate communities. This understanding provides important clues into the political and community entrapment of similarly displaced communities that hinders scholarly understanding. The section concludes with a profile of one displaced KP family that offers a peek into the legacies left behind by those displaced, as reflected in their contributions to the very society that ousted them. This oral history paves the way for a fuller grasp of the family challenges and the complexity of the multilayered transformation that must take place, as discussed later.

Chapter 1: Global phenomenon of internal displacement and Kashmiri Pandit community

Chapter 1 thoughtfully educates the readers on the global crisis of internal displacement and the resulting phenomenon of the IDPs. Distinction is made between those who are forced to flee and seek refuge outside their national borders (Refugees) and those who seek refuge within their own national borders (IDPs). The less understood nature of internal displacement is shown to relate to the sovereignty issue, which assigns the responsibility for care and protection of those who remain internally displaced to the national actors. Among the forces that result in internal displacement, conflict-induced displacement is suggested to leave the most enduring scars, not only affecting those displaced but also spilling onto a whole new generation. The chapter introduces the current debate on internal displacement and unveils the theoretical approach used to dissect and decode the multilayered challenges of those who remain in protracted displacement.

Chapter 2: Context: internal displacement of Kashmiri Pandits

Chapter 2 takes the readers into the historical context of the displacement of the KP community, ousted from their ancestral homes in the Kashmir Valley in 1989. This brings the readers quickly up-to-date on the history of Kashmir and the timeline of events leading up to the displacement of this ethnic minority community. The chapter sheds light on the puzzle as to why the Pandit minority community was specifically targeted when other minorities of the Valley were spared.

Chapter 3: Challenge of researching protracted displacements

Chapter 3 sensitizes the readers to the hidden challenges of investigating the IDP communities, complicated by the protracted nature of such crises. The unveiling of the challenges encountered in investigating the KP community provides the readers with an appreciation of the depth and breadth of the research process requisite for conducting multidimensional and meaningful research in such sensitive communities. Specific hurdles encountered help unfold the political and the community positions embraced by key actors which come to block investigative efforts, relegating these communities to the "hidden population."

14 *Global phenomenon of internal displacement*

Chapter 4: Oral account of one Kashmiri Pandit family and the legacies left behind

This chapter unveils the social, economic, political, and intellectual positions of the KP families before they were ousted. Through an oral account of one displaced family, the chapter exposes the fact that although the Pandits enjoyed good economic and intellectual standing as teachers, doctors, owners of walnut orchards, and well-respected Pandits, as militancy took hold of the Valley, these very people became the key targets of those voicing anti-India sentiments. This glimpse into the pre-displacement provides an appreciation of the legacy left behind by those displaced, as reflected through their roles and contributions to home communities. The adjustment challenges and the predicament of this select family provide a richer grasp of the multilayered transformation that must take place for all those displaced.

Section II: Kashmiri Pandit challenges and the dilemma of return

Section II engages the readers in a methodical and exhaustive analysis of the myriad challenges of the KP community, which during protracted displacement have come to be embedded and disguised in other issues. Given the overlapping nature of the issues, the engagement of the Nested Model (Dugan, 1996) is pursued as the most fitting diagnostic tool to methodically disentangle and analyze myriad issues. Doing so helps illustrate the interconnectivity of the IDP issues of this and similar communities. The dilemma of return, as shaped by the families' own calculus, reflective of displacement-based, lived-in experiences, is illustrated to be the most troubling of the challenges, which keeps the families in cognitive dissonance. Just as the families use their own calculus to support either their reluctance or their longing to return, the officials engage in separate calculations, debating whether to send the families back to where they may be exposed to more harm or keep them on subsidies, indefinitely, in safer communities. Consequently, the issue of return is shown to hold both actors of displacement in perplexity.

Chapter 5: Methodical analysis of Kashmiri Pandit challenges

Chapter 5 details the myriad challenges unfolding in the plight of the displaced Kashmiri families. Given the protracted nature of the KP displacement, the individual issues have come to be meshed and embedded into other issues, thus obscuring a clear understanding of the families' challenges. As the KP policy portfolio has remained woven around the singular identity of "migrants," the Nested Model (Dugan, 1996) has been adapted to systematically untangle the intertwining nature of the psychological, environmental, institutional, and other issues, which underscores the interconnectivity of the IDP issues.

Chapter 6: Moral and political dilemma of return

The dilemma of return is illustrated as the most troubling and the most enduring of the KP challenges. Based on the family's own calculus of displacement-based, lived-in experience, the issue of return holds the KP community in cognitive dissonance, unsure of whether to embrace the new community or prepare for return, rooted in nostalgia for the past. The issue of return is shown to hold both actors in perplexity. This chapter also exposes the moral dilemma of the attempts of the policymakers to encourage Pandits to return, which keeps the lawmakers on the defensive, leading to ambiguous policies, unintended consequences, and the minimizing of the individualized impact of return strategies.

Section III: Policies, assessment, positions, and complexity of policymaking

This section offers a multidimensional understanding of IDP policies, shedding light on the policy assessment by the beneficiaries themselves as weighed through the economic, social, and psychological lens. The positions advanced by members of the civil society are exposed as equally damaging in sending mixed signals that stifle the official strategy. The section provides a unique peek into the individual calculus that informs the families' own assessments of policies, providing exceptional clues into how the displacement experience comes to be weighed as a "blessing in disguise" or a "humiliation to the very being." The displacement policymaking is shown to be constrained by a host of variables, such as an absence of a universally mandated framework, the politics of the host communities, mixed signals sent by the family advocates, resistance from those who never left hometowns, and the constraints implicit in the country's own political agenda. The section concludes by underscoring the complexity of policymaking, thus bringing a unique appreciation to the task of IDP policymaking and the call for context-specific engagement of international actors.

16 *Global phenomenon of internal displacement*

Chapter 7: Kashmiri Pandit families evaluate "migrant" policies

This chapter provides a private peek into the thought process that informs the KP families' own assessment of specific policies, often shaped by their lived-in displacement experience. Given the extended absence from the Valley and with minimal hope of return, the families come to evaluate policies through multiple lenses, such as the economic lens, using a cost-benefit calculation, psychological lens, factoring in the mental tension that comes from dependency, and the social lens that impacts their standing in the host community as the locals view them to be "freeloaders" and "lazy" (Rajput, 2012). The families' assessment of policies, as molded by their context, offers an important perspective on how the IDP experience can be assessed either as a "blessing in disguise" or as a "humiliation to one's very being,"

Chapter 8: How actor positions influence policy outcome

This chapter unveils how the overlapping *positions* (Harre & van Langenhove, 1999) grounded in the interests of the actors influence the formulation as well as the outcome of IDP policies. To facilitate a firm grasp of the actor positions, the rationale behind the toughened positions is decoded and demystified. Often the well-intentioned policies, marred in conflicting positions, backfire, leading to accidental consequences, undermining the policy impact, and thus exaggerating the plight of those displaced. For instance, the elite positions advanced to shelter the policymakers themselves from the multipronged task of seeking long-term solutions are shown to fail, bringing greater harm to those for whom the policies are designed. The positions put forth by the family advocates and members of the civil society are shown to be equally damaging, inadvertently sending mixed signals to the policymakers, thus stifling the official strategy.

Chapter 9: Complexity of IDP policymaking

This chapter takes the readers behind the policies and into the thought processes of policymakers. The IDP policymaking in general and the KP policymaking specifically remains constrained by a host of variables. Obstacles to displacement policies include an absence of a universally mandated framework, the changing missions of the locally based NGOs and members of civil society, the evolving needs of the families, pressure from the host communities, demands of equity from those who never left the hometowns, and the constraints implicit in the country's own politics that are dictated by national agenda. A thoughtful understanding of such bottlenecks not only brings out nuances in an understanding of those who must endure displacement but also brings to bear an appreciation of the national policymaking process and the possible and thus context-specific engagement of international actors in situations of internal displacement.

Section IV: Understanding Kashmiri Pandit displacement through a comparative lens: journeying into Azerbaijan, Georgia, Serbia, and Sudan

While recognizing and respecting that the challenges and the issues unfolded in the preceding chapters are unique to the culture and the nature of the hardships that the KP families encounter, however, in order to gain a deeper appreciation of the predicament of the KP families, issues of similarly displaced communities across the globe are analyzed with the same care and rigor, and put through the Nested Model (Dugan, 1996). Not only does this effort help situate the displacement of this minority community in the context of similarly displaced communities, but more importantly, this indulgence represents a unique effort in displacement literature to zero in on the "core IDP experience." The tracing of commonalities and individualities across communities offers possible clues into societal and policy solutions. The journeying into these communities begins by demystifying the challenges of those who remain displaced after more than two decades in Azerbaijan, where close to half a million people remain displaced from homes in the Nagorno-Karabakh region. Next, the discussion focuses on Georgia, where the families have been ousted from ancestral homes in Abkhazia and South Ossetia, the two regions that became the target of secessionist influence in the early 1990s. This is followed by examining the challenges of those who remain displaced in Serbia, having fled the ethnic tensions in Kosovo in 1999. The comparative journey of displacement concludes with an examination of the plight of the Darfur families displaced in Sudan, where the violence in Darfur that erupted in 2003 displaced more than three million people. Given their prolonged physical absence from home communities and the consequent loss of many rights, these and other displaced families live with minimal hope of return and, in many cases, nonacceptance by their host communities.

18 *Global phenomenon of internal displacement*

Chapter 10: Azerbaijan: displaced from Nagorno-Karabakh

The journeying into similarly displaced communities begins by exploring the challenges of those displaced in Azerbaijan, where close to half a million people remain displaced, after more than two decades, from their homes in the Nagorno-Karabakh region. Issues of the Azeri families are analyzed with the same care and rigor as those of the Kashmir Pandit families and put through the Nested Model.

Chapter 11: Georgia: displaced from Abkhazia and South Ossetia

The journey continues with an examination of those who have remained displaced for over two decades within Georgia, ousted from their ancestral homes in Abkhazia and South Ossetia, the regions that became the target of secessionist influence in the early 1990s.

Chapter 12: Serbia: displaced from Kosovo (ethnic Serbs)

Next, the readers are educated about the challenges of those who remain displaced in Serbia, having fled the ethnic tensions in Kosovo that took place in 1999 and now falling under the rubric of IDPs of Serbia.

Chapter 13: Sudan: displaced from Darfur

The comparative journey of displacement concludes with a robust examination of the plight of the Darfur families displaced in Sudan, where the violence that erupted in 2003 had displaced more than three million people.

Section V: Findings, best practices, and moving forward

Chapter 14: Findings, best practices, and moving forward

Capitalizing on the multipronged research conducted to methodically investigate the protracted displacement of the KP community (Rajput, 2012) and the rigorous commitment to investigating issues of similarly displaced communities across the globe, this revelatory chapter empowers readers across disciplines with over a dozen findings and best practices. The findings are appropriately labeled "Kashmir Specific" and "Broad-Based" findings. Best practices and the lessons learned in the context of the KP displacement are deemed highly relevant in the context of the displaced communities of Azerbaijan, Georgia, Serbia, and Sudan, and are presented in an effort to trace and identify the "core IDP experience and challenges," thus elevating the phenomenon of internal displacement beyond the concerns of the independent nations. The findings are expected to open a platform for multilevel public debate, where the underlying motivations and roles of NGOs, roles of national and international actors, politics and intricacies of IDP/Host dynamics, moral hazards of actor positions, and meaning of return can be better understood to guide the context-sensitive societal and policy solutions. In the context of extended internal displacement, the findings are expected to elevate the issues of IDPs from a matter of humanitarian gesture to one of self-reliance. This holistic understanding offers possible approaches to durable solutions for those who remain in protracted displacements around the globe and raises the likelihood of the formulation of effective displacement policies.

The following chapter will bring the readers quickly up-to-date on the history of Kashmir and the timeline of events leading up to the displacement of the KP minority community from the Kashmir Valley. Additionally, the rich content, unfolded in the remaining chapters, is expected to help seal a painful void created by the absence of scholarly work, exposing the hidden challenges of the KP as well as the similarly displaced communities across the globe.

References

ACCORD. (2010, January). *Conflict induced displacement: The Pandits of Kashmir.* Mount Edgecombe, South Africa: African Centre for the Constructive Resolution of Disputes, 31–37.

Avruch, K. (2011). *Context and pretext in conflict resolution: Culture, identity, power, and practice.* Boulder, CO: Paradigm.

BBC. (May 6, 2015). *Record number internally displaced by conflict.* London, UK: British Broadcasting Corporation.

Darfur genocide, 2003 – Present. (n.d.). *World without genocide.* Retrieved from www.worldwithoutgenocide.org

Dugan, M. (1996, Summer). A nested theory of conflict. *Women in Leadership, 1*(1), 9–20.

Evans, A. (2002). A departure from history: Kashmiri Pandits, 1990–2001. *Contemporary South Asia, 11*(1), 19–37.

Festinger, L. (1957). *A theory of cognitive dissonance.* Stanford, CA: Stanford University Press.

Harré, R., & van Langenhove, L. (1999). *Positioning theory: Moral contexts of intentional action.* Malden, MA: Blackwell.

IDMC. (2014). *Azerbaijan: After more than 20 years, IDPs still urgently need policies to support full integration*. Geneva, Switzerland: Internal Displacement Monitoring Center. Retrieved from www.internal-displacement.org

IDMC. (2016). *Grid 2016. Global report on internal displacement*. Internal Displacement Monitoring Center.

Moza, R. (January 30, 2012). *Why Kashmiri Pandits may never return to Kashmir*. Kafila Organization, Online Blog. Retrieved from: https://kafila.online/2012/01/30/why-kashmiri-pandits-may-never-return-to-kashmir-raju-moza/.

Oberoi, S. (2004). Chapter 8: Ethnic separatism and insurgency in Kashmir. In S. P. Limaye, M. Malik, & R.G. Wirsing (Eds.), *Religious Radicalism and Security in South Asia* (pp. 171–191), 20, Honolulu, HI: Asia-Pacific Center for Security Studies.

OCHA. (January 22, 2015). *The forgotten millions*. Geneva, Switzerland: United Nations Office for the Coordination of Human Affairs.

Prendergast, J. (1997). The nested paradigm of conflict foci: The case of Ethiopia. In J. P. Lederach (Ed.), *Building peace* (pp. 161–169). Washington, DC: United States Institute of Peace.

Rajput, S. (2012). *The displacement of the Kashmiri Pandits: Dynamics of policies and perspectives of policymakers, host communities and the internally displaced persons* (Doctoral dissertation). Fairfax, VA: George Mason University. ISBN: 9781267843333.

Sekhawat, S. (2009). Conflict induced displacement: The Pandits of Kashmir. *Conflict Trends, 2009*(4), 31–37.

Tilly, C. (2005). *Identities, boundaries and social ties*. Boulder, CO: Paradigm Publishers.

UNHCR. (2004). *Guiding principles on internal displacement* (OCHA/IDP/2004/01). Retrieved from www.unocha.org

2 Context
Internal displacement of Kashmiri Pandits

Introduction

This chapter situates the displacement of the Kashmiri Pandit (KP) community in the geographical, historical, social, and political contexts, and sheds light on the puzzle as to why the Pandit minority community was specifically targeted in 1989 when other minority communities of the Kashmir Valley (Valley) had been spared. The displacement of the KPs, treated under the rubric of "sensitive issues" and within the context of the wider "Kashmir problem" (Rajput, 2012), has kept researchers from investigating the triggers as well as the aftermath of this community's displacement. Consequently, now nearly three decades, the painful absence of scholarly examination into this displacement has resulted in story-bound literature that has largely shed light on random, one-sided, and often contradictory explanations of this community's displacement. Additionally, the official positioning of the crisis as a "temporary disturbance" and thus the dubbing of the exodus to a "voluntary migration" (Rajput, 2012) has downplayed the need for any investigation of this crisis as an issue of forced eviction and thus a humanitarian emergency.

Geographical and historical context

Kashmir is a region in south central Asia and forms the northern region of the Indian subcontinent around the Himalayan range. From 16th century, the entire region of Kashmir came to be ruled by the Moguls, Afghans, Sikhs, and the British. Under the British Empire in India (British Raj), Kashmir enjoyed an autonomous status, remaining as one of the several "princely states" within the subcontinent. Soon after the end of the British rule and the consequent partition of India, in 1947, the Kashmir region came to be claimed by both the independent country of India and the newly created nation of Pakistan. Subsequently, the region was incorporated in India, and since that time, the region has become a constant focus of wars over territorial issues between the three neighboring countries of India, Pakistan, and China as well as ethnic, religious, and political disputes over and within the India-administered area of Kashmir. Among the border disputes and the internal conflicts have been three wars between India and Pakistan, a war between India and China, and the forced displacement of the Hindu minority community from the Valley in 1989.

The entire Kashmir region occupies an area equal to 86,000 square miles. Thirty-five percent of the total area falls into the sovereign nation of Pakistan, which has come to be known as the *Azad Kashmir* or Free Kashmir. A smaller portion, roughly 17,000 square miles of the region, remains uninhabited and falls into China, and is known as *Aksai Chin*. The remainder of the area, around 45 percent of the region, the eastern portion, which was officially combined with India, is known as the state of *Jammu and Kashmir* (J&K) and includes Jammu, the Valley, Ladakh, and Siachen (Wolpert, 2010). The J&K state contains several industries, among which are wool and walnut furniture, many of which were owned and operated by the now displaced KP families.

Ethnic and religious context

The 500 years of Muslim rule in Kashmir, from the 12th to the 16th centuries, had led to the conversion of the majority of the population in Kashmir to Islam, leaving only a small population of the Hindu Pandits (Kaw, 2004). In the mid-1980s, the Indian authorities became suspicious of the presence of the "Islamic guerillas" in the Valley, understood to be waging a separatist war dubbed as an indigenous "freedom struggle" (Gill, 2003). This was understood to have left the state of J&K in the "grip of a vicious movement of Islamist extremist terrorism" (Gill, 2003). The anti-India campaign led by the Muslim majority and supported by certain militant groups is said to have taken hold of the Valley in 1989. Groups campaigning for self-determination of the Kashmiri people fought against the Indian government rule. The Indian forces are said to have responded to the outbreak with firings and curfews.

At the time of displacement, the Valley's population was comprised of 94 percent Muslims, and the remaining population was made up of the various ethnic minorities, of which the largest in number (around 5 percent) were the KPs. In addition to the KPs, the Valley's ethnic mosaic cut across and overlapped other minority identities, such as the *Gujjars*, members of the nomadic community; the *Hanjiis*, the artisans of Kashmir; and the *Sikhs*, the entrepreneurs and the supporters of the majority rule. Over the decades of Islamic rule, many of these minority groups had adopted Islam and blended well within the Muslim majority community. However, professing a different faith, the KP minority community was perceived to be representing Indian presence in the Valley and became the specific target of the anti-India movement, whereas other minorities were spared. "Many were threatened, abducted and killed, and those who were forced to flee, formed the pool of 250,000" (ACCORD, 2010) displaced persons, officially dubbed as "migrants" (Rajput, 2012).

Social and political context

The overarching economic, legal, and political framework accessible to the KPs within the Valley had generally compromised their safety and security, even during peaceful times, resulting in *Structural Violence*, as stemming from the "inequitable distribution of resources" (Galtung, 1969). Given the heightened militancy, it became even more difficult for this minority community to protect itself, thus triggering the community's exodus from the Valley. After 27 years of displacement, the officials have often described this displaced community as having become scapegoats of the "[J&K] state government's mismanagement, triggered by a separatist phenomenon" (Rajput, 2012). While almost all Pandits had left the Valley at the height of the 1989 militancy crisis, roughly 2 percent of the KP population had remained in the Valley, clinging to the emotional bonds to "motherland," avoiding uprooting their families, and fearing the phenomenon of "starting over" (Rajput, 2012).

KP standing before displacement

Regardless of the numerous explanations advanced to explain the displacement of this community, the mass exodus of the families had cracked the very soul of an otherwise robust and cohesive community of Hindus, Muslims, and other ethnicities that had prevailed and thrived for multiple centuries, constituting an exemplary mosaic of the Valley. Although the general safety and security of the KPs were generally overlooked, as hard working residents of the Valley, the Pandit families enjoyed good economic and intellectual standing, and made contributions to the community as teachers, doctors, owners of walnut orchards, and caretakers of farm animals. The families enjoyed recognition as "well-respected Pandits" of the Valley and received reverence from the larger community.

Plight of those who fled

Those who initially fled during the height of militancy escaped to safer areas in the nearby towns, such as Jammu, a Hindu-dominated town, within 200 kilometers of their ancestral homes in the Valley. After two decades, some families continue to reside either in the "Migrant Township" (Rajput, 2012) or in the makeshift settlements in the outskirts of Jammu, near the neighboring town of Nagrota, which were hurriedly put together at the height of the exodus. As the KP displacement came to take on the features of a protracted displacement, the families began to secure more economically viable solutions by moving farther toward the metropolitan cities, such as to Delhi and the surrounding cities, thus ensuring economic and long-term survivability.

Conclusion

There have been several attempts at the national and state levels to encourage the families to return to the Valley. However, few have returned, and very few families as a unit have returned, fearing the reemergence of the situation that forced them out in the first place and the expected retaliation from those who never left. However, those who feel compelled to remain outside of Kashmir show resiliency and have thrived in the metropolitan cities due to better education and becoming a more resourceful community. These families are adapting well to making their lives outside of Kashmir.

The following chapter sensitizes the readers to the hidden challenges of investigating the conflict-induced displaced communities, complicated by the unresolved conflicts, resulting in the protracted nature of such crisis. The unveiling of the challenges encountered in investigating the KP community convey an appreciation of the depth and breadth of the research process required for conducting meaningful research into such sensitive and often hidden communities, which are multiplying across the globe.

References

ACCORD. (2010, January). *Conflict induced displacement: The Pandits of Kashmir*. Mount Edgecombe, South Africa: African Centre for the Constructive Resolution of Disputes, 31–37.

Galtung, J. (1969). Violence, peace, and peace research. *Journal of Peace Research*, 6(3), 167–191.

Gill, K. (2003). *The Kashmiri Pandits: An ethnic cleansing the world forgot*. South Asia Terrorism Portal and Institute for Conflict Management.

Kaw, M. (2004). *Kashmir & its people: Studies in the evolution of Kashmir society*. New Delhi, India: APH Publishing House.

Rajput, S. (2012). *The displacement of the Kashmiri Pandits: Dynamics of policies and perspectives of policymakers, host communities and the Internally Displaced Persons* (Doctoral dissertation). Fairfax, VA: George Mason University.

Wolpert, S. (2010). *Continued conflict or cooperating? India and Pakistan*. Berkeley, CA: University of California Press.

3 Challenge of researching protracted displacements

Introduction

This chapter aims to sensitize the readers to the challenges and hurdles often encountered in the investigation of conflict-induced displacements, which become further exaggerated by the protracted nature of such crisis. An in-depth understanding of the challenges encountered in the investigation of the Kashmiri Pandit (KP) community (Rajput, 2012) conveys an appreciation of the depth and breadth of the effort required to gain meaningful understanding of similarly sensitive communities across the globe. Specific hurdles presented here help unfold the political and the community positions embraced by key actors, in the interest of keeping the populations protected from further harm as well as to manage the anxiety of the larger population. However, those very interests and positions come to restrict the investigators' access to these communities, often relegating these communities into hidden populations.

Need for systematic investigation

Given the rising magnitude of the IDP crisis and the number of displacements that are becoming more and more prolonged, extending into several decades, it is crucial that research into the protracted communities be taken seriously and conducted systematically. Such an effort requires exploring multidimensional aspects of displacement. Only such a conversant understanding affords the credibility needed to inform the context-driven approaches in pursuit of durable solutions for the rebuilding of these communities. However, given the caveats attached to the "internal" nature of such displacement, the research into these communities remains a challenge.

Researching protracted displacements 27

The triangulated methodology employed to investigate the displacement of the KP community has unfolded clear as well as subtle clues that often stifle the investigation of displaced communities. Oral histories compiled by engaging with the families of the "migrant" camps, located in the suburbs of Jammu and Delhi, further suggest that the sociopolitical order of the host communities also comes to stifle the researcher's goals of acquiring a well-versed understanding of such communities.

Challenge of investigating KP displacement

The documenting of the challenges encountered in the KP investigation furthers an understanding as to why those who remain in internal displacement remain less understood or often misunderstood. As drawn from the myriad challenges confronted during the KP enquiry (Rajput, 2012), the specific challenges can be understood as residing within the four broad categories: namely, *access to the families, actor positions, contextual factors*, and an *absence of international actors*. These challenges are detailed in the following.

Access to the families

1   Distrust and suspicion
    Once forcibly evicted by their own community members and sometimes community leaders and neighbors, and thus ousted from their ancestral homes, the families quickly become distrustful and suspicious of all those around them, including their neighbors, local leaders, and members of their own government. Such was the case with the KP families, now residing in Jammu township as well as those now residing in the outskirts of Delhi. The distrust and suspicion that the KP families came to embrace were quickly transferred to the members of their new communities, where the families were confronted with strangers. Consequently, members of the host community also came to view the KP community as "sneaky and clever" (Rajput, 2012). In general, the beliefs of doubt and misgiving come to impact the families views of the media and the aid agencies as well as the motives of those who may be trying to investigate their circumstances only to make it possible to improve their situation. These feelings of cynicism and doubt, embraced by those displaced, pose problems for researchers who rely on collecting data through person-to-person contacts or through door-to-door inquiries.

2   Stigmatized populations

The damaging impact of having been forcibly evicted from their own communities goes farther and much deeper than simply the harboring of feelings of distrust and suspicion. An often-irreversible injury stems from the psychological stigma of shame and humiliation that comes from having lost one's "identity, mooring and the sense of belonging" (Korostelina, 2007), consequent to having been ousted from own homes. As these populations shun themselves and become "hidden because of stigma" (Liamputtong & Ezzy, 2005), investigative methods, such as random sampling and face-to-face interviews, become impractical, rendering these communities inaccessible for research purposes. The stigma of shame and humiliation also keeps them out of the official registration process, further reducing the resources available to the researchers.

3   Locating the families

Often, the IDP camps are located in the outskirts of the metropolitan cities, away from the mainstream communities (Annex 1). This is largely due to the housing policy, dictated from the top, that allocates government-arranged township-like settlements for the families. Such forms of housing have been provided to the KPs in the outskirts of Delhi and Jammu. These township-like arrangements are intended to keep the IDP and the Host communities apart for various reasons, such as to minimize the tensions between the two groups, avoid overcrowding of schools, or simply relieve the families of the pressure of adapting to the larger community. Often, keeping the displaced families in these dedicated and compartmentalized communities is explained as a "strategy to help the families maintain their own culture, keep ties to ancestral roots and provide for cohesiveness and bonding with other displaced families," as is claimed in the case of the families displaced from the Nagorno-Karabakh region of Azerbaijan, to be detailed in a later chapter (IDMC, 2014).

However, the IDP housing policy dictated by such strategies not only makes it difficult for the displaced families to access the needed resources but also makes it difficult for the researchers and aid workers to access these communities. Arriving at these difficult-to-access suburbs becomes an arduous undertaking, given the limitations of the city transport. Often, locating the camps successfully requires over a dozen inquiries from the local post offices, law enforcement agencies, vendors, passersby, and elderly residents of the area (Rajput, 2012). Additionally, securing access to these camps often requires the prior approval of the camp coordinator, as in the case of the camps for the KP families. Given their government-arranged accommodations, these families remain confined to their own communities with negligible local interaction. As a result, inquiries from the local residents of the area about the whereabouts of these families often result in unhelpful guidance. Consequently, despite extensive enquiry and footwork, the researchers' efforts often stall and pose delays in successfully locating the families, thus exaggerating the challenges of investigative enquiry.

4   Local resistance

In several cases where the families come to transition into somewhat integrated communities, given their prolonged time in displacement, members of their host communities either become immune to the plight of those displaced or are simply unaware of the continued predicament of the families. Consequently, members of these local communities perceive the efforts of those researching these communities as a frivolous endeavor and withhold useful information from the investigators. Those who may be aware of the crisis also begin to see it as an outdated event and begin to think of those who were displaced as having moved on, making such comments to the researcher as "there are no Kashmiris here, no one knows where they went" (Rajput, 2012). Such resistance from the locals impedes meaningful research into the internally displaced communities.

Actor positions

In addition to the difficulties of locating and accessing the families, there are hurdles that stem from the narratives and the positions embraced by the elite, which also stifle the investigative inquiry.

Official narratives

The manner in which the initial crisis of internal displacement is recognized by the elite comes to either facilitate or obstruct the subsequent research into the displaced communities. The explanation of the KP exodus as an outcome of "voluntary migration" of people into "more stable societies" (Rajput, 2012) was intended to keep the public anxiety under control and to project the displacement as a matter of routine movement. Such strategies are employed by the officials to avoid any possibility of attracting the attention of the larger population, including the media and the research community. Positioning of the KP event in such terms had helped to shape this crisis into a non-event, thus eliminating any need of investigation or intervention into the event.

Culture of keeping silent

The displacement of the KP community has also been shrouded in what is communicated to be the larger "Kashmir problem" that dates back to the historical partition of the country in 1947 between India and Pakistan. The larger problem, understood to be a sensitive issue of political domain, has led the mainstream population to remain silent about the KP exodus. The locally operating humanitarian agencies, taking the lead from the larger population, also remain silent on issues of the politically sensitive nature. Agencies, whose primary function is to provide public information on different aspects of the Indian society, have also come to view the KP displacement as an issue that transpired a long time ago and one that has been "resolved," thus discouraging the researchers' efforts.

Contextual factors

The researchers' efforts are further compounded by the contextual factors of specific displacements, as explained in the following.

Living spaces inside the camps

Shamed by the interior of their living spaces, the displaced families are often reluctant to speak to researchers and other outsiders in and around their living spaces. This adds to the challenges of the research effort as this may mean conducting face-to-face enquiries with family members on the roadside; in secluded areas; or in the midst of traffic, pollution, and animal noises in order to respect the families' privacy issues and be sufficiently away from their living quarters. This is often done with the hope of discouraging the researcher's effort, hoping that the researcher would leave even before the process begins. On the other hand, the families that do manage to come to terms with their living spaces and willingly expose their living conditions to the researchers often become distracted by the interior of their camps during the interview process, drawing the researcher's attention to the broken windows, the leaking ceilings, and the flickering light bulbs inside their camps, allowing themselves to derail and stifle the interview process (Rajput, 2012).

Protracted nature of displacement

In the case of protracted displacements which endure for multiple decades, such as that of the Darfur families of Sudan who have been displaced for over 15 years; the KPs, displaced since 1989; or the Abkhazia and the South Ossetia families of Georgia, displaced since 1990, the researcher's challenges further multiply. Over the prolonged time in exile, as the families evolve and enlarge, and as the policies come and disappear, in the interest of survivability, some families find themselves having become somewhat integrated with the host population. Over time and in the "process of acculturation" (Berry, Kim, Minde, & Mok, 1987), displaced families take on or adopt similar attire, behavioral traits, and social customs as those of the host communities. Such assimilation makes the researcher's task harder to identify and distinguish the displaced families, adding another complexity to the research effort. Additionally, given the prolonged displacement, the mission of the local NGOs that proliferated at the height of the crisis also subside or fold into other activities, taking on new missions, thus making it difficult for the researcher to count on their support toward a meaningful understanding or the tracing of the whereabouts of the communities.

Absence of international actors

In addition to these challenges, specifically encountered in the investigation of the KP community, the most prominent feature of internal displacement that assigns the responsibility for the care and protection of those displaced to the national actors accounts for one of the major hurdles in any investigative inquiry on internal displacement, as explained in the following.

Independent nations, as sovereign states, often discourage external actors from intervening in a nation's internal matters, such as the conflict-induced crisis that triggers displacement of communities from their hometowns. In some cases, the national leaders block agencies from providing any form of support, humanitarian or technical, in response to such emergencies. National actors maintain that external intervention and involvement can threaten the country's stability, a position taken up by Sudan's leadership, which expelled several international aid organizations in 2009 (*The Guardian*, 2009). The absence of international actors on the ground allows latitude for the national actors to manage and address the crisis as appropriate from their own perspective, which often results in the framing as well as the misdiagnosing of the crisis in ambiguous and mysterious ways.

Further, as the vertical nature of the top-down societies dictates (Galtung, 1996), the national-level responsibility further comes to be delegated to the lower, local- and the municipal-level agencies. In the interest of keeping the concerns of the local communities at bay, the local leaders also come to downplay the crisis and communicate the arrival of newcomers in secretive ways. This results in stifling the exploration efforts that can help unfold the complex nature of the issues that confront the displaced communities. Consequently, cases of internal displacements that remain less understood and less investigated remain plentiful across the globe. However, in cases where the national leadership is deemed incompetent or unwilling to address the crisis of internal displacement, international actors are called to respond to such crisis. The presence of international actors and agencies, and the information and the guidance available through these external partners becomes a valuable resource for the much-needed research effort required to understand the true nature of the issues that confront the lives of the displaced families remaining in exile within their own borders.

Conclusion

This itemization of the challenges encountered into the investigation of the KP community helps explain how despite the displacement of this community, now exposed to be one of the most traumatic realities for this community, only a handful of studies have focused on this displacement. This has resulted in painful gaps in the research-based, holistic understanding of this community. Going forward, it is hoped that the challenges documented here can help shape future investigative processes, aiming to secure meaningful insights into the communities that remain unexplored, mislabeled, and misunderstood.

The following chapter offers a multidimensional lens into the profiles of the displaced families, showcasing the incredible insights that often become available only after having conquered the research hurdles exposed in this chapter. The family profile, constructed from one family's own narratives, provides an appreciation of the family values and the legacy left behind by those displaced as reflected through their roles and contributions to the very societies that ousted them. The adjustment challenges and the predicament of the family in their transitional journey is also noted in order to provide a richer grasp of the multilayered transformation that must take place by all families once displaced.

References

Berry, J., Kim, U., Minde, T., & Mok, D. (1987). Comparative studies of acculturative stress. *International Migration Review, 21*(3), 49–51.

Galtung, J. (1996). *Peace by peaceful means: Peace and conflict, development and civilization.* London, UK: Sage Publications.

IDMC. (2014). *Azerbaijan: After more than 20 years, IDPs still urgently need policies to support full integration.* Internal Displacement Monitoring Center. Retrieved from www.internal-displacement.org

Korostelina, K. (2007). *Social identity and conflict: Structures, dynamics, and implications.* London, UK: Palgrave Macmillan.

Liamputtong, P., & Ezzy, D. (2005). Researching the vulnerable. In *Qualitative research methods.* Oxford, UK: Oxford University Press.

Rajput, S. (2012). *The displacement of the Kashmiri Pandits: Dynamics of policies and perspectives of policymakers, host communities and the internally displaced persons* (Doctoral dissertation). Fairfax, VA: George Mason University. ISBN: 9781267843333.

The Guardian. (March 5, 2009). Sudan aid agencies expelled. *The Guardian.* Retrieved from www.theguardian.com

4 Oral account of one Kashmiri Pandit family and the legacies left behind

Introduction

This chapter offers a rare glimpse into the life of one displaced Kashmiri Pandit (KP) family, shedding light on the social, economic, and intellectual standing of the Pandits in the Kashmir Valley (Valley) before displacement. This robust understanding of one family provides a deep appreciation of the legacies left behind by those displaced, as reflected through their values, roles, and contributions, to the very communities that came to oust them. The profile of this Pandit family is constructed (Rajput, 2012) from an oral account of the family's own recall, as narrated after 26 years of exile, recalling the life before displacement. Once displaced, the families cling to such lived-in experiences that once formed an integral part of their community's fabric and fulfilled their own sense of purpose (Korostelina, 2007), and which, during displacement, forms their nostalgia for the past. The continued predicament of the family in its transitional adjustment further affords a richer grasp of the multilayered transformation that must take place for all those displaced, elsewhere.

Ecological setting

An understanding of the ecological settings of the families' lived-in spaces provides direct as well as subtle clues into the mind-set behind key IDP policies of housing, jobs, and schools. In search of safety, this one Pandit family fled from its ancestral home in the Valley in 1990 and now lives in Jammu's "Mini Township for Migrants (Jagti)" camps. The family's flat inside the Jagti Camp is located in the army cantonment area of Jammu, away from the mainstream community, and guarded by the government's security forces. The township, inaugurated in 2011, is designed to house approximately 4,000 "Pandit migrant" families. The government-funded township provides the families with "transitional accommodations" with the understanding that the families will eventually return to the Valley. The township serves as a self-standing community with schools, hospitals, community halls, and temples within the camp premises, some built by the KP community itself. The small-scale shops and the schools in the compound are run and managed by the KP families. This peek into the ecological settings of the KP family suggests that the officials view the KP displacement as more than temporary.

Family profile

An understanding of the family profile offers clues into the crossover influence and the generational spillover of a protracted displacement. Anticipating a quick return to the Valley, the Pandit family initially travelled to the closest town of Udhampur, 55 kilometers away from their Valley home. The family recalls that their sons were between the ages of three and five when the family fled the Valley. Having moved multiple times since displacement, the family now lives 298 kilometers from their hometown. Before displacement, the family spent their entire life in the Valley, within a society that was made up of 94% Muslims, the remaining population making up the various ethnic minorities, which is in contrast to their current Jammu community, comprising 65% Hindus and 31% Muslims (India Census District Profiles, 2001).

This family of eight members represents overlapping generations, ranging from the grandfather, aged 75, to a 5-month-old grandson. Prior to displacement, the family enjoyed a thriving walnut tree business in the Valley. After the initial hurdles of settling in the new place, the life for the family resumed in the "transition" quarters. The parents prioritized their responsibility of educating their children and later arranging their children's wedding and witnessing the birth of their grandchild, all as part of the life in the migrant camp. The senior grandparents live in their own government-allotted flat, separate from the flats allotted to their sons and their families. The son was able to secure an accounting job within Jammu's mainstream community and travels by bus from the township to his workplace. The grandmother tends to the household chores and remains mindful of keeping the larger Pandit family together as a unit. The junior family has two sons and the parents take great pride in the fact that even in the midst of camp life, they were able to avail of the government's education benefits extended to the displaced KP families and were later able to send their sons to college. Consequently, the parents have now raised two self-supporting sons.

Keeping with the Kashmiri tradition of marrying within the community, the family located a suitable match for their son within the "migrant" community. The family regrets that after displacement, this time-honored tradition of marrying within community was disrupted for many KP families, leading to "intercaste marriages." The bride's family, also a KP, was displaced from the Valley in 1990 and now lives in a nearby "migrant" camp in Nagrota. The wedding of the son, with KP traditions, took place in the "Migrant Township" community hall built by the displaced families themselves. The five-month-old son of the junior Pandits, the great grandson of the Pandit grandfather, was also born in the "migrant" township.

The many achievements of this Pandit family, despite being uprooted from their hometown, and the many moves while in exile reflect their priorities and values. These traits hint to the principles that may have contributed to the robust fabric of their hometown community of the Valley.

Circumstances of eviction

The oral account, as narrated by the Pandit grandfather, provides a multidimensional lens into the circumstances that led to the family's eviction. The grandfather recalls that the family had become accustomed to hearing the daily broadcasts of threats relayed from the mosques, telling the Pandit families to leave the Valley. Similar to other families, the Pandit family understood these threats to be of a "temporary nature" and prayed for calm in the Valley. The grandfather recalled that the state officials, too, had warned the Pandits that "not every house could be protected from militants" (Rajput, 2012). In the interest of protecting the family from harm and having reached the "threshold of tolerance and the constant mental abuse inflicted by the militants," the grandfather felt compelled to flee the Valley with the family. The family's hope of returning to the Valley within a short time meant "not wanting to go too far" so as to make it "easier to return."

The grandfather explained that organizing the funds for travel was not a problem for the families as "no Pandits were short of money, they were not poor, all Pandits worked hard, earned good income and were respected by the locals" (family grandfather, personal communication, July 18, 2011). The Pandit family reaped good income from their 25 walnut trees. They regret leaving behind a three-story house and many of their belongings as "only so much could fit on the rented truck." The grandfather also laments not being able to bring the family animals and the trees that had ensured the family's security in the Valley. The painful thought of having to leave the family animals behind, when the house became surrounded by the militants, haunts the family even after 26 years. The grandfather lamented the irony of having a storage full of grains but not being able to bring even a small portion for the uncertain journey ahead.

The grandmother explained that regardless of how the militants had treated the families, as Pandits they "never picked up weapons and had no training in using them," embracing the belief that "the one who cannot save another's life also has no right to destroy one's life." The family's belief in the "order of God" meant "refraining oneself from harming another" and reflected their coping skills and optimism in the midst of the humiliation of eviction. The family praised their neighbors: the "majority of our friends were Muslims, the elders respected us, we celebrated festivals together." The mother recalled that their neighbors were friendly and that only some evil spirit had forced them to unite with the militants against the Hindu families. The grandfather recalled, "we were very comfortable in our homes but then people started disappearing, one of our neighbors had received a note [leave Kashmir or prepare to be killed], with the increasing frequency of such threats, we became terrified." All 20 people living in the Pandit household were able to flee. The grandfather regretted that "not all families were able to leave, as about 200,000 people's houses were burnt, with people killed inside the houses in front of their eyes." He understood that the militancy had inflicted pain on all parts of the community, remarking that "all suffered from militancy, militancy was bad for the entire community and not only for those who were forced out." Those who stayed behind and those who had evicted the families also suffered.

The personal account of this Pandit family offers valuable insights into how the family's sense of dignity and regard for human life may have contributed to their resiliency and coping skills while in displacement.

Adjustment challenges

An understanding of the displacement challenges offers subtle clues into the extent of the IDP policies and provides an appreciation of how the families come to manage their displacement.

Journeying into the unknown

After the initial 55-kilometer journey from the Valley, the family arrived in the neighboring town of Udhampur and faced their first challenge of finding a place to rent. In the interest of maintaining the family unity, the Pandit family declined offers from those who were unwilling to accept all the family members in their homes. Having come only a short distance from their hometown, from here on, the family's life, as well their cultural and personal identity, was to transform drastically from here on.

Transitional accommodations

The Pandit family does not consider the Jagti Township home, but they breathe a sigh of relief to be in a community shared by other displaced Pandit families. Living in a self-sustaining community has reduced the family's need to explore the city beyond the camp premises. They recall early encounters with the locals of Jammu as those of humiliation and shame, being ridiculed by the locals as "cowardly" in not being strong enough to face their aggressors in the Valley and for having abandoned their homes instead of protecting them. The family also suspected a hint of jealousy from the locals as the Pandit families had become the recipients of housing, education, and other day-to-day basic necessities. This sentiment by the members of the host community reflects a sense of "relative deprivation" (Gurr, 1970), where the members of the host community came to entertain a perception of being treated unfairly in comparison with the KP families, who were seen as unqualified and unworthy of the state benefits that they were now receiving, simply on the account of having "abandoned" their homes in the Valley.

Navigating the bureaucracies

Initially, the Pandit family struggled to navigate the bureaucracies required for registration, filing for compensation, and finding school for their children. In addition, the hotter climate of Jammu, which the family was not accustomed to, made it difficult for them to travel from place to place with two infants to find the administrative offices. The presence of security forces outside the township also created the perception that perhaps it was unsafe to venture outside the township. Consequently, the family became housebound, spending extended hours within the confines of their allotted apartment. At the same time, they experienced the anxiety of being separated from other relatives and worried about the property and the animals left unattended in the Valley and in the hands of the militants.

Loss of permanence

As the Pandit family had never left the Valley before, their sentiments of casting their future as "dark" and as an "end of life" signaled a threat to their sense of permanence and cultural values in seeing their lives being removed from places "connected to past social practices" (Avruch, 2011). Their loss of permanence is exaggerated by the positions that the officials maintain regarding the nature of the KP displacement. The families of the township are continually reminded of their transitional accommodations which they will need to vacate once the "situation normalizes" in the Valley. This has created a moral dilemma for the Pandit family in whether to accept and embrace their new community wholeheartedly or to treat the members of their new community as chance encounters. This loss of permanence is reflected in the family's day-to-day vocabulary, with the frequent use of words such as "for now," "in the meantime," and "if God wishes."

Dilution of identity

The Pandit grandfather laments the loss of his personal identity, remarking:

> I was a well-known Pandit, in a society in which the ratio of Pandits was much less, the Valley people knew me as respectable Pandit Ji, the owner of the walnut orchards. Now I am nothing, a migrant of Flat X, a Pandit among 2,500 other Pandits. I have lost my social standing.
> (Grandfather of the Jagti camp, July 20, 2011)

The family feels stigmatized by the need to carry a government-issued plastic identity card. They are also slowly coming to terms with seeing themselves not as a joint Pandit family with sons, daughter-in-law, and grandchildren all living in one unit but as a nuclear family, with the family members living on separate floors. This new family structure represents a demotion of the family values. The Pandit family is mindful of the gradual weakening of their ethnic identity over the years in exile, as reflected in their changed lifestyle, such as a preference for floral prints replacing their Kashmiri embroidered outfits and the authentic Kashmiri cuisine slowly being replaced by assorted cuisine. The daily ritual of worship is also slowly being replaced by the prerecorded prayers in lieu of the live singing and chanting.

This account of the adjustment challenges of this KP family helps shed light on the gradual dilution of a community's identity and a diminished sense of permanence when displacement comes to take on the features of a permanent exile.

Future outlook

Although fragmented into separate family units, living on different floors of the camp, the family elders strive to uphold the joint family values by eating and worshiping together, and they are comforted by the fact that at least, they are able to live in the same building. All family members from the 75-year-old grandfather to the 5-month-old infant are part of a happy household, where everyone is neatly dressed, elders are treated with respect, and the youngsters are showered with love in return. The family regrets that the traditions of the Valley are slowly disappearing, with most people having been forced to live outside of the Valley, but the family recalls life in the Valley as a happy time of their lives, with the memories of the "majestic mountains" behind their home etched into their memory. In the midst of living with the fear of being uprooted again, the hope of a brighter future for their children keeps the Pandit family motivated; the grandmother remarks, "I have no land or trees anymore, but I have educated my children, now I want better for my children."

In the absence of KP policies that would guarantee the family's safe return, the idea of return looks bleak to the Pandit family. The grandfather admits that although the township is made up of the KP families, it is missing the love and affection experienced in the Valley, thereby perpetuating the nostalgia for the past, for how things used to be. At the same time, the family is mindful of the challenges of returning to the Valley, among which, fearing retaliation from those who never left and those who evicted them, their perpetrators.

Conclusion

Although the oral history of this one Pandit family is based on contextual settings, shaped by the lived-in experiences of their time before and while in displacement, the adjustment challenges and the continued predicament faced by them stands as a powerful testimony to the predicament of similarly displaced families. Accordingly, the account of this one family serves as a treasured guide to the societal and the policy solutions required for understanding as well as the rebuilding of similarly displaced communities.

Given the protracted nature of the KP displacement, the individual issues of KP community have come to be meshed, embedded, and disguised into other issues, thus obscuring a clear understanding of the families' challenges. The next chapter takes on the task of decoding and demystifying the challenges of protracted displacement.

References

Avruch, K. (2011). *Context and pretext in conflict resolution: Culture, identity, power, and practice*. Boulder, CO: Paradigm Publishers.

Gurr, T. (1970). *Why men rebel*. Princeton, NY: Princeton University Press.

India Census District Profiles. (2001). *Basic data sheet*. Office of the Registrar General & Census Commissioner, India. Retrieved from www.censusindia.gov

Korostelina, K. (2007). *Social identity and conflict: Structures, dynamics, and implications*. London, UK: Palgrave Macmillan.

Rajput, S. (2012). *The displacement of the Kashmiri Pandits: Dynamics of policies and perspectives of policymakers, host communities and the internally displaced persons* (Doctoral dissertation). Fairfax, VA: George Mason University. ISBN: 9781267843333.

Section II
Kashmiri Pandit challenges and the dilemma of return

5 Methodical analysis of Kashmiri Pandit challenges

Introduction

This chapter empowers the readers with a systematic and a holistic understanding of the myriad challenges encountered by the Kashmiri Pandit (KP) families as they remain in exile. Given the protracted nature of violence-triggered displacements, the family issues come to be meshed and entrenched into other issues, thus obscuring a clear understanding of the true nature of the specific and critical issues. Additionally, as the KP policy portfolio has remained interlaced around the singular "migrant" identity and in order to zero-in on the outstanding challenges, it is crucial to carefully grasp the intertwining nature of the psychological, environmental, institutional, and technical issues of this community. Doing so is hoped to open space for family-oriented solutions required to address the unsettled issues of this community.

Complex nature of displacement

Internal displacement is a complex issue as it touches and adversely magnifies all aspects of those displaced, such as the individual, social, economic, legal, and political. In addition to the challenge of "starting over," groups displaced by conflicts continue to be marginalized by members of their host communities, sidelined by the authorities, and forgotten by members of their own home communities. For administrative and simplification purposes, once displaced, the myriad issues encountered by the families come to be nested into a homogenous and monolithic grouping, such as the issues of the *displaced*, the *migrants*, the *guests*, or the *outsiders*, which, in some communities, are referred to as issues of the *temporaries*. However, failure to recognize the specific as well as the holistic nature of the challenges of those displaced is risky, as it often leads to misguided solutions, policies, and practices that frequently produce unintended consequences and at best yield only minimal effect.

Analytical framework

A meaningful and a systematic understanding of the overlapping and the interconnecting challenges of the KP families can be unfolded by invoking Maire Dugan's Nested Model. This Model is used in social sciences to analyze conflicts as "something dynamic and organic, and as products of events occurring at various interactive levels" (Dugan, 1996).

In Dugan's framework, a conflict is understood to be contained within a comprehensive system, comprising of four circles (Figure 5.1). The smallest circle is surrounded by progressively larger and larger circles; therefore, the immediate issue confronting the conflict is traced to the larger systemic aspects. Of immediate relevance is a *specific* issue, which, in turn, is embedded in elements of a *relational* nature, the latter being entrenched within the larger context, comprising *subsystem* issues, and the deepest and the underlying layer of the conflict system resides into the society-wide issues of a *structural* nature (Prendergast, 1997). This interconnectivity of issues creates a complex web, suggesting the importance of addressing the underlying elements of any given crisis within this conflict system. The specific challenges of the KP community, framed around these four components of Dugan's Model, are deciphered in the following paragraphs.

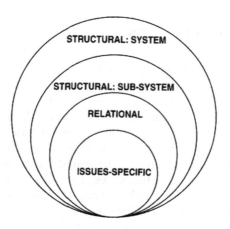

Figure 5.1 Dugan's Nested Model of Conflict.

Specific issue

The specific issue of the comprehensive conflict system addresses the immediate issue faced by the people. The displaced KP families point to a host of reasons for having fled the Kashmir Valley (Annex 2): "lack of security," "militant threats," "targeted killings," "increased Islamization," "to escape suffering," and "to protect their children" (Rajput, 2012). However, once having left the Valley, the unprecedented experience of having become *homeless* surfaced as the shared sentiment of the community among all those who fled. The following section illustrates the phenomenon of homelessness in the context of KP displacement.

After nearly three decades, the trauma of having become homeless continues to surface as the key issue for all KP families. The thought of having been evicted from community carries a psychological shock of having lost homes and been made to face an uncertain future. For many families, the *specific* challenge was moving out of their ancestral homes and villages for the first time, which amounted to the phenomenon of homelessness. Having lived in Kashmir their entire lives, the ancestral roots and emotional ties for these families resided in the Valley. The tremor of being uprooted from familial homes was humiliating, and that shameful experience of the day of their departure is now unanimously shared by all those who fled (Rajput, 2012) and forms this community's "collective memory" (Kagee & Del Soto, 2003, p. 27).

Consequent to the loss of home, the "loss of identity," "belonging," and the shattered "self-esteem" (Korostelina, 2007) has now become this community's "chosen trauma" (Volkan, 2001). Volkan explains "chosen trauma" as "the shared mental representation of a massive trauma that the group's ancestors suffered at the hand of an enemy" (pp. 79–97). As the idea of culture is "connected to past social practices as well as to the ongoing phenomenon" (Avruch, 2011), the losses from eviction extended beyond the loss of the physical structure of their homes and stretched into the loss of their cultural, social, economic, and political fabric. The family's separation from the Valley led to a feeling of "helplessness," "victimization," and an "end of future." The departure signaled a break in the family unity, shared values, and common life experiences. The uprooting of the entire community had also placed the families in jeopardy by increasing their insecurity, triggering challenges of a psychological nature. The lost homes have now come to represent a metaphorical space of personal attachment, exemplar of "how things used to be" (Achieng, 2003, p. 17).

This stigma of homelessness was experienced by all those who fled, even those who had initially stayed with their own relatives. Ironically, those who succeeded in finding safety at their nearby relatives' homes within and around Jammu felt a deeper sense of humiliation as the relatives' unrestricted hospitality that extended at the beginning of their unexpected arrival gradually came to an end as their stay prolonged. The overextended welcome signals sent the families back to the humbling task of searching for accommodations in the "migrant" camps, which by this time had "maxed out," having exceeded their capacity, additionally some camps had also closed their registration process. These "migrant camps" were no longer registering new arrivals. Having been absent from the initial camp registration process, the fate of these families now lay in limbo as, by this time, they had lost their right to accommodations on account of being displaced from the Valley and feared the likely loss of several displacement-related future rights.

Regardless of the contested and much-debated forces, and the personal reasons that had triggered the individual family's departure, the exact date and time of that traumatic day of their departure carries a permanent scar in their psyche, having played out several times during their time in exile. The national response to provide this community with "township"-like settlements away from their ancestral homes has not alleviated this stigma of homelessness and isolation, even after nearly three decades. As residents of two-room tenements in multistory complexes in Jammu, the families continue to lament the loss of their own family homes and feel deprived of a shelter to call their own.

In the context of the Nested Model, the policy response that called for exclusively addressing the immediate needs of the families, through "transitional accommodations" in compartmentalized communities, has left the deeper concerns of these families, which lie beneath the *specific* issue, unaddressed and thus unresolved.

Relational issues

Relational issues of Dugan's conflict system relate to the *fit of the people* into a society. The fit of the people can be stifled by the psychological, physical, cultural, environmental, and other bottlenecks that people face upon arrival into a new community. Subsequent to displacement, finding one's place, and making meaning in a new society involves an intricate and complex IDP/Host dynamic, which can at times produce strong intercommunity bonds or bring the societies to their lowest points. Establishing such bonds requires that members of both societies have an opportunity as well as a willingness to reach out to each other (Allport, 1954). "Pressure from the local population, which typically must shelter a large number of outsiders" with limited resources (Calderon, 2010) poses additional challenges for the IDPs, thus impacting their fit into the new society. Such pressures often result in "resentment, hostility, and mistreatment of IDPs" (Duncan, 2005; Ferris, 2011). In a similar vein, coping with the demands of their new society had put tremendous pressure on the KP families, such as when having to navigate the bureaucracies of the big cities, required for filing for compensation, and searching for schools and medical facilities (Rajput, 2012). The fit of the displaced KP families into their new surroundings was hampered by a number of such factors, as explained in the following.

1 Changed social milieu

The forced eviction of the KPs had broken up their families, fragmented their social and traditional ties, and had halted employment, education, and marriage aspirations for many. Once in new communities, the KP families were tasked with the primary mission of restoring the family life and dignity. However, arrival into new cities suddenly exposed the KP families to the changed social milieu. As respectable traders and *Pandits* of the Valley, the metamorphosis into "anonymous migrants" was humiliating for the families. In the Valley, the KP families practiced the *joint family system*, a family arrangement, which allows and even requires multiple generations to live within the same household; the pre-militancy environment of the Valley was described by the families as "very comfortable," infused with "brotherly love" and "respect" (Rajput, 2012). For some families, the most troubling loss from eviction was not their displacement per se but the exposure to societies with "diluted" moral values, the nuclear family structures, and intercaste marriages that were being embraced by members of their host communities. Such social dislocation continues to trouble the families after multiple decades of having resided in these new communities outside of the Valley.

2 Complex Host/IDP dynamics

Soon after the locals and the hosting communities of Delhi and Jammu realized that the stay of the KPs was more than temporary, they "developed an antipathy towards this community" (Pandita, 2013, p. 134), which came to be reflected in the local landlords trying to "control every aspect of their lives and ridiculing them for their language and pronunciation" (Pandita, 2013, p. 134). The host communities that had initially sympathized and welcomed the KP families, such as the Delhi's *Baapu Dham* community, eventually began to reassert control over their resources, such as their wedding halls (*barat ghar*), which were being used to house the KP families. The locals made a case to push the KP families out of their community, asking them to vacate those halls (Rajput, 2015). The local community's appeal to their own state officials led to relocating the families to the outskirts of Delhi and other cities hosting the KP families. The seemingly well-intended policy response to house this community in distant "migrant quarters," to manage and solve the IDP/Host rift, led to the stifling of the IDP/Host relations. The government-arranged, self-sustaining townships created a cultural divide, with both groups limiting their social interactions to members of own in-groups. This has exaggerated the "IDP/Host differences" (Korostelina, 2007), reinforcing the stereotypical images that each community holds of the other. Such sentiments are now reflected in the KP community's characterization of the locals as "materialistic," "arrogant," and "uncaring," and similarly in the characterization of the KP community members by the locals as "greedy," "lazy," and "troublemakers" (Rajput, 2012).

Additionally, these townships, meant to provide a safe haven and a close-to-home-like experience for the families, have incurred the moral hazard of robbing the KP community of the resources needed to rebuild their society and hampering the IDP/Host relations. Ironically, those residing in the "townships," despite being surrounded by their own group members in close-knit communities with temples, schools, and shops, talk of "living in a vacuum without political space and rights" and miss the "brotherly love of their Valley neighbors" (Rajput, 2012), conveying the nostalgia for the past.

However, a limited number of families that could afford and made their way into Delhi's mainstream communities found opportunities to reach out to their wider community by volunteering and tutoring, strengthening both the local and IDP communities with mutual gains. Their visibility among the locals has boosted their own self-esteem, helping them to reclaim some of the lost social worth. While forging new friendships, these families also preserved their own culture (Berry, Kim, Minde, & Mok, 1987), consequently faring better socially and economically. This outcome contrasts with their Jammu counterparts, who continue to find themselves in community and political entrapment as their social orientations remain dictated by the "township"-style housing model. Consequently, the larger segment of the KP families remain traumatized by the loss of their ancestral homes, which has hampered their adjustment into new communities, thus impacting the *relational* aspects of displacement in the context of the second circle of Dugan's Nested Model.

Subsystem issues

In reference to the Nested Model, *subsystem* issues of a conflict system are issues that reside below the surface of the specific or the immediate issue, such as one's upbringing and the family value system. Although the subsystem issues lie below the surface, such matters percolate and come to impact issues at other levels. For instance, in the context of Colombia's displacement, Zea explains that "social upbringing impacts how people identify themselves and others in their society" (Zea, 2011); such values come to govern intergroup behavior, an important aspect of IDP/Host dynamic.

The relational aspects of the KP families, that is, their ability to communicate and integrate into the local Jammu and Delhi communities, were initially hampered by the differences in language and education, and the social values embraced by the KP families as residents of the Valley. An exposure to what the KPs perceived to be a "mixed society" diluted with "intercaste marriages" discouraged the families' participation in the issues of their wider community. Nonparticipation reduced the families' mobility, which eventually came to be perceived by the locals as the antisocial behavior of the "migrants." These perceptions reflected in the host families' statements about the KP families as those who "stay to themselves" and as those who "do not know how to mix." Consequently, this cultural divide led to the formation of two parallel societies. Over the years, the subsequent changes that have come to be introduced into the KP family system continue to clash with KP family values, such as the "advent of inter-caste marriages, which is becoming widespread and rampant, and increasingly becoming a rule rather than the exception" (Moza, 2012, p. 1). The families feel a sense of the "disintegration of their community through loss of language, culture and traditions, fearing that their community may be disappearing" (Raina, 2010). A sense of demotion of their traditional and cherished family values remains the most troubling aspect of the community's ouster from the Valley.

Structural issues

Structural issues of Dugan's comprehensive conflict system pertain to issues that reside at the deepest layer of the Nested Model (Figure 5.1). These issues have the potency to percolate and become a breeding ground for grievances, such as societies' rules and regulations that govern access, compliance, and behavior of members of the society. Depending on "how these structures are set up, they can either facilitate the lives of those [displaced] or oppress them further" (Galtung, 1969), as illustrated in the following.

1. Historic marginalization

   The overarching economic, legal, and political framework accessible to the KP families exaggerated their many challenges not only after being displaced but also before displacement, even as contributing and hardworking residents of the Valley. As an ethnic minority community, the safety and security of the KP families was generally compromised, even during stable and peaceful times, making them the victims of systemic violence. Galtung explains that the social structures that systematically prevent people from meeting their basic needs signify a form of violence: namely, *structural violence* (Galtung, 1969, pp. 167–191). As the everyday policies disproportionally excluded the KP families, it became even more difficult for them to protect themselves at the height of militancy. The families suggest that in the absence of any repercussions, for the denigration of the KP community, "public chanting of death threats and celebrations of their killings" became the norm in the Valley. Later, consequent to displacement, the structural policies and systems of the Valley impacted the families' access to community, property, and their right to return, thus impacting their socioeconomic and political standing for all time to come.

2. Obstacles to restoring livelihood

   Securing adequate means of livelihood and access to services is challenging for any displaced community, and the added pressure that spills into the host communities manifest as heightened tensions between the two groups, thus exaggerating their *relational* aspects within Dugan's Nested Model. To alleviate such pressures, the policy package had included the "temporary use" of the shops that were made available for the use of the KP families in the local communities. This has given the families an opportunity to regain their economic well-being. However, the use of these shops poses a predicament for the families, as the regulatory compliances placed on these shops stifles the family's growth. Having designated the crisis of KP displacement a "temporary disturbance," the temporary use of the shops is aligned with the official "positioning of the crisis" (Harré & van Langenhove, 1999) and is likely to stay in effect until the families are able to return to the Valley. This positioning has allowed the officials to maintain an ad hoc mix of temporal policies in lieu of more sustainable and durable solutions. Accordingly, the government retains the ownership of these shops, restricting the shopkeepers from making any structural changes, such as extending the shop parameters in periods of high sales or even protecting the shops through insurance mechanisms as doing so would convey ownership of the shops by the shopkeepers versus ownership in the hands of the officials.

3 Misfortunes of labeling

Access to services available to the KP community has been further blocked, given their "migrant" label. At the outset of KP displacement, such labeling was likely a convenient mechanism to identify members of this community for the purpose of allocating housing, shops, and subsidies. However, such labeling has led to the most enduring of the socioeconomic misfortunes for the families. The locals of Delhi and Jammu use these very markers as their own tools to exert "power relations," to "draw boundaries," and to "dictate rules for inclusion/exclusion" (Tilly, 2005, pp. 6–7). Even after multiple decades, social inclusion remains an issue for the KP community, as the locals discourage the inclusion of "migrants" and "outsiders" in their community-wide matters. The KP families continue to be perceived as "temporaries" by the locals. The identity descriptions that refer to the KPs as "temporaries" and "outsiders" have also made them easy targets of local law enforcement officials, such as the police, who feel at liberty to harass them for not knowing the rules of the society. In addition, as the "migrant" label suggests a voluntary departure from the Valley as opposed to "forced eviction," such labeling has also offended the KP community as the authorities have not held their perpetrators accountable for the community's forced eviction.

4 Irony of prolonged absence

In addition to the challenges associated with policies and procedures of new communities, KP families have also been challenged by meeting their responsibilities back in the Valley. Given their prolonged absence from the Valley, the families have lost many rights that had been afforded to them as residents of the Valley.

However, the state bureaucracies continued to hold the families accountable for their pending obligations owed to the Valley officials. Those with properties left behind found themselves obligated to care for their homes and pay property taxes, ironically, for houses from which they were forcibly removed and perhaps may never return to. An official of a KP-based advocacy group commented that "back in the Valley, when census people come and knock on the Pandit family's door, and month after month no one answers the door, their name gets removed from the voter registration, and they lose all rights" (AIKS, personal communication, August 22, 2011).

In general, the KP policies flowing from the top-down structures of the host as well as the hometown communities have stifled the adjustment of the families, thus exaggerating the challenges consequent to displacement. Having partially overcome the initial hurdles of displacement, the families remain challenged after almost three decades as they continue to make meaning of their lives, rebuilding from their spiritual past, amidst an uncertain future.

Conclusion

The engagement of Dugan's Nested Model has facilitated a sound grasp of the specific challenges of the KP families. It has become clear that the sociopolitical and socioeconomic impact of past policies, and the continued predicament of the community has remained grounded in the initial positioning of this displacement as a matter of random disturbance. More importantly, the displacement of KPs and the overall phenomenon of global displacement can now be understood as a multidimensional and comprehensive system in contrast to a standalone crisis resulting in mono-dimensional issues of an economic, political, or social nature. The interconnectivity of the IDP issues, as exhibited by those displaced from the Valley, suggests a similar comprehensive approach to community building.

The following chapter unfolds what remains the most enduring of the KP challenges: namely, the issue of return. The chapter exposes the truth that the dilemma of return not only holds the KP community captive in total perplexity, but the issue of return also keeps the officials on the defensive, often leading to ambiguous and contradictory positions and policies, sending mixed signals to the families, and thus prolonging the family anxiety and undermining the official return strategies.

References

Achieng, R. (2003, February). *Home here and home there: Janus-faced IDPs in Kenya*. Conference report presented at Researching Internal Displacement: State of the Art, Trondheim, Norway.

Allport, G. (1954). *The nature of prejudice*. Cambridge, MA: Perseus Books.

Avruch, K. (2011). *Context and pretext in conflict resolution: Culture, identity, power, and practice*. Boulder, CO: Paradigm Publishers.

Berry, J., Kim, U., Minde, T., & Mok, D. (1987). Comparative studies of acculturative stress. *International Migration Review, 21*(3), 49–51.

Calderon, M. (2010). Internal displacement: Recent history, visions for the action ahead. *ISP Collection*. Paper 838. SIT Graduate Institute, Brattleboro, VT.

Duncan, C. (2005). Unwelcome guests: Relations between internally displaced persons and their hosts in North Sulawesi, Indonesia. *Journal of Refugee Studies, 18*(1), 25.

Dugan, M. (Summer, 1996). A nested theory of conflict. *Women in Leadership, 1*(1), 9–20.

Ferris, E. (Ed.). (2011). *Resolving internal displacement: Prospects for local integration*. Washington, DC: The Brookings Institution.

Galtung, J. (1969). Violence, peace, and peace research. *Journal of Peace Research, 6*(3), 167–191.

Harré, R., & van Langenhove, L. (1999). *Positioning theory: Moral contexts of intentional action*. Maiden, MA: Blackwell.

Kagee, A., & Del Soto, A. (2003, February). *Taking issue with trauma. Researching Internal Displacement*. Conference report presented at Researching Internal Displacement: State of the Art, Trondheim, Norway, p. 27.

Korostelina, K. (2007). *Social identity and conflict: Structures, dynamics, and implications*. New York, NY: Palgrave Macmillan.

Moza, R. (2012, January 30). *Why Kashmiri Pandits may never return to Kashmir*. Kafila Organization, Online Blog. Retrieved from: https://kafila.online/2012/01/30/why-kashmiri-pandits-may-never-return-to-kashmir-raju-moza/.

Pandita, R. (2013). *Our moon has blood clots: The exodus of the Kashmiri Pandits*. New Delhi, India: Random House.

Prendergast, J. (1997). The nested paradigm of conflict Foci: The case of Ethiopia. In J. P. Lederach (Ed.), *Building peace* (pp. 161–169). Washington, DC: United States Institute of Peace.

Raina, D. (2010). Kashmiri Pandits facing "extinction". Retrieved from http://in.groups.yahoo.com/group/kashmiripandit/message/76

Rajput, S. (2012). *The displacement of the Kashmiri Pandits: Dynamics of policies and perspectives of policymakers, host communities and the internally displaced persons* (Doctoral dissertation). Fairfax, VA: George Mason University. ISBN: 9781267843333.

Rajput, S. (2015). Chapter 3: Internal displacement of Kashmiri Pandits. In S. Kukreja (Ed.), *State, society, and minorities in South and Southeast Asia*, New York, NY: Lexington Books, p. 62.

Tilly, C. (2005). *Identities, boundaries and social ties*. Boulder, CO: Paradigm Publishers.

Volkan, V. (2001). Transgenerational transmissions and chosen traumas: An aspect of large-group identity, *Group Analysis, 34*(1), 79–97. Center for the Study of Mind and Human Interaction: University of Virginia, Charlottesville Virginia.

Zea, J. (2011). Internal displacement: Violence, public and socioeconomic policy in Colombia. *The Michigan Journal of Public Affairs, 8*. University of Michigan, Ann Arbor.

6 Moral and political dilemma of return

Introduction

This chapter offers multi-perspectivity and a nuanced understanding of the meaning of *return*, the issue that comes to be the most troubling and the most enduring consequence of having been forcibly evicted. In circumstances of violence-trigged displacements, the return to family homes becomes an issue of a political, legal, economic, and moral nature. Consequently, the issue of return remains one of the most conflicting and painful issues for the Kashmiri Pandit (KP) families, with strong disconnect between all parties: namely, the officials, members of the host communities, the family advocates, the families that yearn to return, and the families that are repelled by the idea of return. Just as the families use their own calculus to support either their longing to return or their reluctance to return, the officials, guided by their own *positions*, also engage in separate calculations of their own, debating whether to send the families back to where they may be exposed to more harm or to sustain them with temporary housing, education, and other subsidies, thus delaying pondering over this contentious issue. While bringing to clarity the meaning of return, this chapter exposes the dilemma of return not only for those displaced but also exposes the gaps in the efforts of the policymakers to encourage the KP families to return.

Meaning of return

Over the course of extended displacement, the idea of returning to their hometowns takes on different meanings for different communities. For some, return to ancestral communities relates to the moral obligation and the loyalty owed to one's hometowns, which can only be fulfilled by returning to care for the farms and the lands that once nourished them and their young ones. This duty-bound view of return fulfills the family's need to contribute to the prosperity of their ancestral towns. For others, return means a return of one's identity, dignity, self-esteem, which can only be regained by returning to one's inherited bonds (Korostelina, 2007); in some cases, return means the right to "die with honor in the beloved homeland" (personal communication, July 18, 2011, KP grandfather), signifying a symbolic meaning of return. For the KP families, the issue of return also comes to be weighed through the KP "culture," with the idea of norms, values, right, and good (Avruch, 2011).

Similarly, reluctance to return carries different meanings for different families. For some, unwillingness or resistance to return, despite the incentives offered by lawmakers, means loyalty and gratitude owed to the new communities that housed them in their hour of need. For others, unwillingness to return means a moral obligation to protect the family from further harm, which may be inflicted by returning to a community that ridiculed their own identity and belonging.

Regardless of the meanings ascribed, the concept of return includes aspects much broader than the mere physical return to their hometowns. Return calls for securing economic restructuring, rehabilitation, the revival of social life, and reconciliation with the perpetrators (Aker, Celik, Kurban, Unalan, & Yukseker, 2006). Over the course of prolonged displacement, the idea, as well as the prospect of return, is shaped by the individual positions held by various parties and actors, as explained in the following paragraphs.

Family perspective

The Kashmir Valley (Valley) community that had nurtured and cared for the young and the elderly, and that had provided meaning for the life's experiences and a vision for the future has now become a "collective memory" (Volkan, 1997) for those who remain in displacement. Sekhawat (2009, p. 132) has noted the "social and the cultural deterioration of this community as the major fallout of the KP displacement, as the entire community was forced into exile." Most families had lost what may have taken them generations to build, leading them to mark their departure as a lifetime of suffering where "everything had changed ... as [they] were leaving for good" (Pandita, 2013, p. 98). After nearly three decades, those who long to return admit that the "social fabric of their society has changed forever, the seeds of mistrust have been sown, and that society can never be trusted again, and lament of having been robbed of the right to die in homeland" (personal communication, July 27, 2011, Delhi-based KP).

For the KP families, the issue of return has been highly emotional and most debatable, even among the members of the same family. They remain conflicted, some with unwavering desire for return; others disgusted by that very idea; and still others in *cognitive dissonance* (Festinger, 1957), feeling reluctant to wholeheartedly commit to a "mixed society" and equally reluctant to return to a society that "humiliated their identity." Embedded in this psychological tension rests their moral dilemma of choosing between the two options, when both options entail less than the desired outcome. The KP families contemplate whether embracing a new society will reflect disloyalty to the homeland or whether returning to the beloved homeland will expose their families to even bigger dangers. Consequently, for many families, the thought of return keeps them in perpetual uncertainty, unsure of whether to embrace the new communities or to prepare for return, thus producing a dilemma of return.

Family voices express dilemma of return

> I left on March 6, 1990, when they [militants] began targeted killings. They killed our leaders and killed my neighbor. I thought I will go back, but on January 26, 1994, on my Republic Day, they burnt my house in celebration of our eviction. Here [in Delhi] we live in our own [KP migrant] community. I am alright but my kids do not understand where I am from, they do not understand Kashmiri language. I have no identity here. I will go back if I can have my own Kashmir but they grabbed my property, where will I go?
> (personal communication, August 14, 2011, Delhi camp)

The issue of return has been troubling for not only those who were forced out of their ancestral homes but also the whole new generation born and brought up in the "migrant" camps as they remain clueless of the societies that their elders talk about and the legacy they left behind. Oberoi (2004) suggests that "this youth is growing up and is being shaped by a different culture and may not feel as part of Kashmir," reinforcing the families' argument for "no return."

Additionally, the youth that now makes up a part of the fragmented Valley community is growing up in largely a homogenous community, under a single religion and a singular political ideology, and in an absence of diversity. This youth of the Valley also remains clueless of the KPs that once made up their own community (Oberoi 2004). It is uncertain how this group will receive the KP families on their return to the Valley.

The following paragraphs shed light on the calculus used by the KP families that informs their views on return.

Reluctance to return

Despite the assessment of the "migrant" policies by the KP families as "politically motivated," "humiliating," or even "irrelevant" (Annex 2), these families admit that they are not interested in returning to the Valley as their children are now working, and additionally, they no longer have a place to return as they have abandoned their Valley homes and properties. These families point to a broken community in the Valley, where they fear being tormented again on return, and they would rather forget everything that happened. The shopkeepers are reluctant to return, fearing that they will "barely survive [in Valley] and opine that conditions will never change enough for them to want to return" (personal communication, July 26, 2011, Delhi-based KP shopkeeper). Oberoi (2004) notes the demographic shift of the Valley, which now represents a population consisting of 99.5% Muslims, with scant minorities. Consequently, with the departure of the KPs, the economic landscape of the Valley has also changed, with the KP-owned shops now bolted and out of business. The parents of schoolchildren fear that their kids "will be forced to go to Islamic schools." These families point to a host of reasons, such as "destruction of religious symbols," "increased Islamization," "lack of security," "challenge of starting over," and the "changed social fabric of the Valley" (Rajput, 2012) as their reasons for having ruled out their return. These families position their displacement as an irreversible crisis that has permanently damaged their community. In their reluctance to return, the KP families convey their own *positions* on how their society, the schools, and the places of worship should be run and conducted (Avruch, 2011). These positions have come to form the family's decision-making, which spills into other aspects of their displacement.

Evans (2002) suggests that those who have left the Valley are integrating well with their fellow Indians, anticipating that the migrating families will assimilate with the communities around them, which may lead to the disappearance of the hometown culture. Congruent with this sentiment, a number of families have sold their houses in the Valley and have stopped participating in Kashmir's elections, signaling their lack of interest in the future of Kashmir. Having come to terms with their displacement, some families see their exodus as a blessing in disguise. Implicit in the families' "no return" rationale is a sense of the families' new and "acquired identity," which reflects the families' personal abilities, merit, and preferences (Korostelina, 2007, pp. 79–81), signaling some level of acceptance of the values and norms of their newly adopted, and what some families may have come to view as their permanent, communities.

Longing to return

Whereas most KP families, recalling the bitter memories of the circumstances of their eviction, have ruled out their return, a smaller number hold dearly to the possibility of return, while some families find themselves in a state of perplexity. The latter group endures the mental tension of entertaining both the possibility of return as well as the possibility of staying in new communities. Those "unceasingly yearning to return cling onto memories and the hope that someday peace will prevail and they will be able to return home" (Dhingra, 2013). However, the families that hold dearly to the possibility of return make it known to policymakers that their return is conditional on multiple factors, such as a separate homeland, the removal of security forces, the restoring of normalcy, and a concrete blueprint with timelines that guarantee their dignified and sustainable return.

Comparative perspective of Delhi- and Jammu-based KP families

A comparison of the Jammu- and Delhi-based families (Rajput, 2012) that continue to reside in the "migrant" camps reveals that whereas all Jammu families have ruled out their return, a handful of Delhi families still hold on to a desire to return. Jammu families convey that they no longer have a place to return to as their houses were torched or that their kids will be forced to go to state-mandated schools. In contrast, the Delhi families that still hold on to a solid desire to return talk about never having adjusted to the hot climate and feeling pressured to live in communities with "low moral values." However, even those with their hearts in Kashmir and a longing to return to the beloved birthplace admit that there is "danger" and "absence of freedom" in the Valley, and fear that the "government may have taken away their last rights of dying in homeland" (personal communication, July 29, 2011, Delhi-based KP family of Rohini camp). Those placing conditions on their return also add the caveat that even if they were to return, they fear that their children may not return with them and remain doubtful whether the Valley society can be trusted again. Regardless of the opportunities available in their respective and distinct communities, both Delhi- and Jammu-based families reminisce about the brotherly love enjoyed in the Valley, which they see as missing in what they describe as "very materialistic communities." This nostalgia for the past serves to perpetuate the families' anxiety.

Host community perspective

There is general agreement between members of host communities of both Delhi and Jammu in how they view the KP families' intentions to return. Members of both communities perceive return of the KP families as "highly unlikely," believing that "seventy-five percent of the KP families have sold their homes in Kashmir Valley and have transferred their money to our [host] communities" (personal communication, July 23, 2011, Jammu business community). Host communities also position the KP families as being weak in not being able to face their perpetrators when they were being forced out and for having abandoned their hometowns. For that reason, the host communities are of the opinion that the families will "never" go back.

Some host families have concluded that after these families left the Valley, their KP "culture had died," and their identity is now lost. In addition, the locals believe that having benefitted from the education, jobs, and housing policies, the families have little incentive to return. The locals suspect that the combination of prolonged exile and the opportunity-driven communities may have fostered new identities in the families, which now factor in the families' own cost-benefit calculus. Such narratives help to reinforce the host families "no return" argument, the host families also view the return policies as "fabrication" simply showcased by lawmakers to keep the KP families from protesting. The perceptions shared by both local communities of Delhi and Jammu are largely aligned with the perspectives of the officials, who declare that "policies are designed to encourage the KP families to return, however, the families are less likely to return, as they are being pressured by their own youth to stay" (personal communication, Delhi official, August 1, 2011).

Policymaker perspective

Policies made in response to the crisis and those that evolved during the period of decades-long displacement remain grounded in the initial *position* (Harré & van Langenhove, 1999) that the policymakers had come to embrace to explain the initial emergency. Consequently, the officials continue to address this displacement as a "temporary crisis," reiterating that the "families must go back." Consistent with this view, the Jammu township families are continually reminded that they are to return "when normalcy happens." In the meantime, the officials strive to take care of the families until they are ready to return.

The disconnect in the families' own positions has also led the officials to advance a mix of ambiguous policies. On the one hand, certain policies are touted as honoring the families' own wishes to return, such as the job and housing incentives offered to those willing to return. Simultaneously, other policies are showcased to facilitate the families' integration into new communities, such as permitting the families to purchase property in their new communities.

Although this dual narrative is likely intended to shield the elite themselves from the urgency of either formulating a robust return policy or providing for full integration into new societies, ironically, the same narratives have been adopted by the KP families to maintain their own shifting and dual positions and narratives.

Capitalizing on the mixed signals, first understanding that the officials want the "families to go back," the KP families feel rightly positioned to demand concrete policies and timelines of their return. Second, sensing that the officials actually doubt whether the families will go back, the families feel rightly positioned to demand full rehabilitation in new communities. This dual narrative and storyline has not only kept the KP families in perplexity but is also keeping the officials on the defensive. The officials are tasked with the dual mandate of addressing the concerns of those who hold dearly to the possibility of return while simultaneously catering to the wishes of those who demand "dignified placement" into new societies.

However, the officials stay alert to the sentiment that given the elapsed time and space, even if older people want to return, they will most likely be pressured by their youth to stay in new communities. Accordingly, the officials expect that contrary to the demands and the "noise" of return by some, the disconnect between the older generation – those with emotional attachment to ancestral roots – and the younger generation – those with forward-looking vision – will eventually coincide, thus facilitating the formulation of concrete and all-encompassing KP policy portfolio.

Conclusion

Similar to other displaced communities around the world, the dilemma of return for the KP community is more than an issue of the generation gap or an issue driven by a simple cost-benefit calculus.

The issue of return is deeply entrenched in the mix of moral and cultural stance of this community as well as in the positions of key actors, which go beyond assigning theoretical explanations or cost implications for the family's continued predicament. The officials remain cognizant that the issue of return has been tricky for both the families and the policymakers. Pending the issue of return, the families continue to confront the dilemma of whether to embrace the new opportunities and help their own family grow or pursue their moral quest to return to the beloved homeland. The government, on the other hand, strives to provide for both rehabilitation in the new communities and to offer incentives to those willing to return.

Not only has the issue of return posed dilemmas for the policymakers and the KP families, but also the overall mix of policies, as guided by the official positioning of this displacement, has often led to accidental consequences and minimal impact of policies, thus exaggerating the predicament of the families. The following chapter offers a private peek into the thought process of how the KP families come to evaluate specific policies, which provide an extraordinary understanding of how displacement experience can come to be weighed as a "blessing in disguise" or a "humiliation to the very being" by different families of the same displaced community.

References

Aker, T., Celik, B., Kurban, D., Unalan, T., & Yukseker, H. (2006). *The problem of internal displacement in Turkey: Assessment and policy proposals.* Istanbul: Turkish Economic and Social Studies Foundation (TESEV).

Avruch, K. (2011). *Context and pretext in conflict resolution: Culture, identity, power, and practice.* Boulder, CO: Paradigm Publishers.

Dhingra, V. (2013, November 23). Leaving past behind, Kashmiris adopt Pathankot as new home. *Hindustan Times.* Retrieved from www.hindustantimes.com

Evans, A. (2002). A departure from history: Kashmiri Pandits, 1990–2001. *Contemporary South Asia, 11*(1), 19–37.

Festinger, L. (1957). *A theory of cognitive dissonance.* Palo Alto, CA: Stanford University Press.

Harré, R., & van Langenhove, L. (1999). *Positioning theory: Moral contexts of intentional action.* Malden, MA: Blackwell.

Korostelina, K. (2007). *Social identity and conflict: Structures, dynamics, and implications.* New York, NY: Palgrave Macmillan.

Oberoi, S. (2004). Chapter 8: Ethnic separatism and insurgency in Kashmir. In S. P. Limaye, M. Malik, & R.G. Wirsing (Eds.), *Religious Radicalism and Security in South Asia* (pp. 171–191), 20, Honolulu, HI: Asia-Pacific Center for Security Studies.

Pandita, R. (2013). *Our moon has blood clots: The exodus of the Kashmiri Pandits*. New Delhi, India: Random House.

Rajput, S. (2012). *The displacement of the Kashmiri Pandits: Dynamics of policies and perspectives of policymakers, host communities and the internally displaced persons* (Doctoral dissertation). Fairfax, VA: George Mason University. ISBN: 9781267843333.

Sekhawat, S. (2009). Conflict induced displacement: The Pandits of Kashmir. *Conflict Trends, 2009*(4), 31–37.

Volkan, V. (1997). *Blood lines: From ethnic pride to ethnic terrorism* (2nd ed.). Boulder, CO: Westview Press Boulder, Colorado.

Section III
Policies, assessment, positions, and complexity of policymaking

7 Kashmiri Pandit families evaluate "migrant" policies

Introduction

This chapter provides an exclusive peek into the thought process and the personal calculus employed by the Kashmiri Pandit (KP) families that come to inform the families' own assessment of "migrant" policies (Annex 2). The beneficiaries' assessment of policies, as shaped by the political and the community settings of their host communities, provides an extraordinary understanding of how the displacement experience can come to be weighed as a "blessing in disguise" or a "humiliation to the very being" by different families of the same community.

During nearly three decades of in-country exile, the families have seen an evolution of the overall policy portfolio promoted to "help alleviate [their] suffering" (personal communication, August 1, 2011, Delhi officials). However, consistently framed around the explanation of the crisis as a "temporary disturbance," the policies have maintained an interim feature "aimed to help the families until they are ready to return" (personal communication, August 1, 2011, Delhi officials). Given their extended physical absence from the Valley with minimal hope of return and the consequent hardships of adjustment into new communities, the families come to evaluate policies through distinct and multiple lenses. The KP families use the economic lens to make the cost-benefit calculation, assessing the costs of starting over in hometown communities as weighed against the financial benefits of remaining in the host metropolitan communities. The families also use the psychological lens that helps them to factor in the mental tension and the humiliation that comes from their dependency on state-provided subsidies. Weighing the policies from a social lens helps the families assess the impact of policies on their standing in their host community, where the families experience being perceived as "free riders," "lazy," or "dishonest" by the locals.

Evolution of KP policies

The officials suggest that the KP policies are grounded in an understanding that, given the "brief uproar" in the Valley in the 1980s, people got frightened and voluntarily moved to safer areas, and needed immediate and temporary shelter (Rajput, 2012). Consequently, the policies that emerged and evolved during the families' time outside the Valley are claimed to be in direct response to the families' needs and demands, and driven by the humanitarian goal of reducing pain of those in distress. The officials clarify the policy intent to "provide short-term relief to families, as people want to go back" (personal communication, Delhi official, August 1, 2011). The current as well as the expired policies are touted to be fulfilling this interim purpose of helping the families feel "comfortable until they are ready to return." However, such positioning has left the families in a predicament, suspicious of whether the policies are truly grounded in benevolent principles or whether the policies could be a sinister plot to keep the families in indefinite transition.

KP policy portfolio

During this extended displacement, a host of KP policies have come to be formulated, revised, and extended; some have been allowed to remain intact and some policies have been terminated. In addition to the financial assistance provided to the families to meet their day-to-day needs, the key policies that form the current policy portfolio focus on housing, children's education, and jobs. The evolution, nature, and the content of these policies are detailed later.

Housing

The KP housing policy has evolved significantly from the provision of the initial makeshift camps, intended to accommodate the families as they were arriving into new communities to the current, dedicated multi-story "township-like settlements" in the outskirts of Jammu. The initial camps were assembled hurriedly with tarp and other roofing materials that proved to be ill-suited for housing the families for multiple decades. During the second decade of displacement, many of these initial camps began to be replaced by the township accommodations: multi-story apartment complexes designed to serve as the self-sustaining communities for the KP families. Located on the outskirts of Jammu, these complexes are solely designated to house the displaced KP families but remain under government control and ownership. Although most families were able to transition to the newer accommodations, some families remain in the initial makeshift camps on the outskirts of Jammu, such as in Nagrota or in the Muthi camps. The most durable housing solution has come in the form of an option extended to the families to purchase the subsidized housing in their host communities. Implicit in this strategy is an acknowledgment by the officials of the high-level uncertainty of the families' return to the Kashmir Valley. Implemented through the Delhi Development Authority, this option has allowed the Delhi-based families to purchase suitable housing to meet their families' growing needs. The bureaucracies, enforced on local home purchasers, have also been lifted for the KP families who wish to purchase housing in designated host communities on the outskirts of Delhi, such as in the active development districts of Dwarka Sector, a distance of 34 kilometers from Delhi, and Rohini, one of the 12 zones administrated by the Delhi municipality.

Education

Although not all KP families came to learn of the education policies while residing in the makeshift camps, education for the school-age children has been available since the arrival of the first families into the new communities. The education initiative, as directed by the "Prime Minister's Special Allocation for the Kashmiri Migrants," has been responsible for providing education for all school-aged children of the KP families. All children have been eligible for up to high-school education, free of charge. Under this policy, "an average of ten children from the KP displaced community have been admitted annually to any of the 1,000 government schools" (personal communication. August 3, 2011, Principal, Kendriya Vidalaya School). The high school graduates have also been eligible for college education within mainstream communities.

Jobs

For the KP families, job-related policies have manifested in several forms. Those who were forced to flee and consequently lost their Valley businesses have been allotted makeshift shops on a "temporary" basis in several cities, now home to the families. The provisional use of the shops has allowed the families to produce and sell their craft, and showcase their skills through a range of services provided for the larger communities. Within a less than one-mile radius of Delhi's Indian National Army Market, over 50 shops can be found in the area designated as "Kashmiri Migrant Market" where the Kashmiri families are able to sell clothes, shoes, and handicraft, and provide medical, beauty, and tailoring services. This has helped the families not only to regain their livelihood but also to reclaim their social capital (Korostelina, 2007) and to recover their standing in the host communities. Additionally, the "job quota" extended by the national authorities has also expedited the recruitment process for the KP youth. The quota allows a limited number of qualified KPs to become part of the local employment industry, bypassing the bureaucracies otherwise imposed during the rigorous recruitment process. In addition to the allocation of the shops and the job quota, the salaries of those who were employed in the government sector before displacement have also been grandfathered and thus protected. This salary protection for those eligible has allowed these families to continue to draw their salaries while in in-country exile, thus freeing their time and talent to volunteer and contribute to the affairs of their host societies.

Family assessment of policies

Most KP families are aware of the various policies that emerged in the early phases of displacement as well as those that followed in subsequent years. They are equally aware of the policies that emerged randomly and disappeared quickly without impact. The evaluation of the policy portfolio has varied from policies being hailed by the families as "beneficial" and sometimes even as a "blessing in disguise" to the policies being understood as having "failed" or been "useless," "politically motivated," or even "humiliating" (Rajput, 2012). The assessment of policies as weighed through the family's own calculus using varied lenses is provided in the following paragraphs (also detailed in Annex 2).

Beneficial policies

The policies most often ranked by the KP families as having been beneficial are largely driven by the family's cost-benefit calculation and thus assessed through the economic lens. Within the overall KP policy portfolio, the policies that the families rank to be most beneficial, relate to housing; initial tools made available to resume livelihood; the provision of education for their youth; and the salary protection for those who were employed in the public sector before displacement. Most families speak of having benefitted from these policies, at least during some phase of their displacement; owing to these policies, some families have come to attribute their displacement experience as a "blessing in disguise."

Delhi-based families praise the tools made available to them as they were beginning their lives in the *Hauz Rani* camps, such as the availability of sewing machines. The provision of the sewing machines allowed the families to resume some level of livelihood. The education policy under the national initiative has been touted by most KP families as having had the greatest impact. The families proudly speak of having educated their children despite living in the "migrant" camps. For many, the accessibility of the schools on the camp premises was a blessing in the early phases of their displacement, when they found themselves overwhelmed, having to navigate the new community and trying to understand the rules of new societies. The families also talk of their youth having benefited from the job quotas that provided employment to the qualified KP youth. Those who were beyond the school years, proudly talk of having benefitted from the temporary use of the shops, provided in select host communities. Those who were working for the government before displacement praise the policy that has kept their pre-displacement Valley salaries protected, which has freed their time and allowed the families to share their expertise to help advance the work of their new host communities.

Failed or useless policies

Depending on the specific needs and composition of the family; the demographics of the communities; and, most importantly, the lens used to assess the policies, the families' assessment of policies differs sharply. Whereas some families see the provisional use of the shops as having enabled them to regain their economic well-being, other families see the same policy as "completely useless." The temporary use of the shops and the Jammu settlements, which were intended to help the families reclaim their sense of security, has often been assessed through the psychological lens and ranked as "failed" or "useless." The families see these policies as standing in contrast to the official stance, which they understand to be to ensure the return of the families to the Valley. Consequently, the township settlements and the allocation of the shops are understood by the families as sinister ploys to keep them in indefinite migrant status, thereby eroding the families' sense of confidence and pride. These families see the policies as a "failure in what the officials really wanted to accomplish for [our] families" (personal communication, family advocate, July 28, 2011). The families are further convinced that since they are not a "vote bank," their priorities have been sidelined, and provisions such as the use of the shops are absent of "well-thought out and deserving policies."

Politically motivated policies

There is widespread disagreement among the KP community on what they understand to be the motives and intent behind the policies. In contrast to the official claims that promote KP policies as grounded in the humanitarian agenda, many families understand the policies to be premature, politically motivated, grounded in the political gains, and "purely meant to garner votes." For instance, the making of the Jammu township has angered many KP families, who perceive the settlement as a "denial of their right to return," and this has rattled the families' confidence in the government's ability to secure their eventual return. In addition, the families also view the return policy to be solely boosting the government's image. They are of the opinion that "the policies to send them back to Kashmir are not viable and only look good on paper" (personal communication, KP advocate, August 5, 2011).

Humiliating policies

Many families point to the policies as being "humiliating to the very being," referring to the initial phases of their displacement, where inaccessibility of the needed support structures forced the families to go from camp to camp in search of accommodations. Additionally, being asked to live in "quarters that were previously used to house the animals at night" (personal communication, July 20, 2011, Jagti camp) was perceived by the KP families as akin to being treated like animals. The experience of having stayed in those quarters in the early days of displacement remains a stigmatizing experience for the families. The overcrowded camps, with bed sheets and women's *sarees* used to separate each family's space, also meant that the young females of the households could not be protected and provided a private space. Even after their move to the township communities, the "migrant" policies that had separated the families from the wider community continue to be seen as humiliating and impact the family's social standing in the wider community.

Conclusion

This assessment of policies by KP families themselves makes it clear that the displacement experience can be interpreted as a blessing in disguise or as a humiliation to the very being, depending on how the families come to interpret and understand the underlying motives and positions of those who are entrusted to make the policies. Although the families are able to assess several policies as having been beneficial, there remains general agreement in the wider KP community that the policies have not been wide enough, have been premature and politically motivated, and involve complex and humiliating procedures to access. The families continue to feel shamed and disgraced by the host communities' perception of the KP community as freeloaders, lazy, and deceitful, given their continued dependence on these policies.

The following chapter helps unveil how the overlapping actor positions, including the signals sent by the families and the family advocates to their lawmakers, come to influence the formulation as well as the outcome of displacement policies, thus complicating the task of formulating displacement policymaking.

References

Korostelina, K. (2007). *Social identity and conflict: Structures, dynamics, and implications.* New York, NY: Palgrave Macmillan.

Rajput, S. (2012). *The displacement of the Kashmiri Pandits: Dynamics of policies and perspectives of policymakers, host communities and the internally displaced persons* (Doctoral dissertation). Fairfax, VA: George Mason University. ISBN: 9781267843333.

8 How actor positions influence policy outcome

Introduction

This chapter offers a nuanced perspective of the actor *positions* (Harré & van Langenhove, 1999), thus shedding light on how the conflicting positions that remain grounded in the interests of the actors themselves often backfire and negate the outcome of good intentions and policies. In circumstances of long-term displacements, the narratives, often advanced to liberate the policymakers from the multi-pronged task of securing durable solutions, often fail, thus negating the impact of otherwise well-intended strategy. Equally damaging for the Kashmiri Pandit (KP) families have been the incompatible positions advanced by their own advocates, and members of the civil society, who, have often shifted their positions based on their own changing priorities. The families themselves remaining in perpetual uncertainty and a permanent void also inadvertently send mixed signals to lawmakers, thus influencing the outcome of key policies. The following sections help demystify the rationale behind the toughened positions embraced by these actors, which come to either stifle the policies or result in premature or ill-suited policies.

Elite positions

This section brings to light the thought process that keeps the officials grounded and loyal to their individual positions despite the marginal effect of certain policies. As was disclosed earlier and is worth repeating here, for nearly three decades, the KP policy portfolio has largely remained a function of the initial positions advanced by the national leaders to address the crisis of displacement as it unfolded in the Kashmir Valley (Valley; Rajput, 2012). The unwavering embrace of these narratives over the years has led to a progression of an ad hoc portfolio of policies, often simply meant to reinforce those explanations. However, these hardened accounts have led to a complex situation for the officials themselves, undermining their own efforts of reducing the families' anxiety about remaining outside home communities.

Two key explanations advanced by the authorities describe the KP displacement as: (i) a temporary disturbance/instability in the Valley and (ii) a voluntary migration (personal communication, August 1, 2011, Delhi official). The following segment is an attempt to decode these explanations and to unfold the policy impact of these conflicting explanations.

KP displacement explained as a temporary disturbance

One of the two explanations used to describe the KP displacement has been an articulation of the crisis as an outcome of a "temporary disturbance triggered by a separatist phenomenon, a temporary takeover by hostile elements" (personal communication, August 1, 2011, Delhi official). Accordingly, the officials expect the families to return to the Valley once the situation normalizes. Operating from this understanding, the officials claim that the formulated policies are in direct proportion and response to the families' immediate and transitional needs for food, clothing, and shelter. Under this goodwill scenario, the KP families are treated as "guests," and the officials strive to ensure the families' safety by keeping them protected in safer communities. Just as the families are being sheltered under the interim policies, the officials too continue to be sheltered from the burdensome task of securing solutions of a more sustainable nature. This has allowed the policy portfolio to remain grounded in treating the KP families as "victims of fear," justifying a temporary fix to alleviate this fear. The explanation that the families' stay is temporary and thus brief has also helped to justify the policy of keeping the KP families in segregated communities, which is anticipated to make it easier for the families to reintegrate back into hometown communities on their eventual return.

These tactics have left the families in a state of perpetual uncertainty, where they are now overly cautious about integrating into new societies and equally fearful about return to what they perceive to be an unsafe and broken hometown community. The segregation of the KP families has also limited the community's access to the resources required to rebuild their community. Entrapped in the positioning of displacement as a temporary event, the families feel locked out of growth prospects, thus extending their precarious socioeconomic standing, which continues to "stigmatize the families as poor and helpless, obstructing their efforts to improve their self-reliance" (Wistrand, 2013).

KP displacement explained as a voluntary migration

In explaining the ouster of the KP community, the officials maintain dual accounts of the community's displacement. Under certain circumstances, the authorities explain this crisis as having been triggered by a "temporary instability," and at other times, the unprecedented departure of this community is explained as the outcome of a "voluntary choice." The authorities contend that as the "Valley environment, became more volatile, people became self-motivated to seek better opportunities and fled of own volition, triggering a mass exodus" (personal communication, August 1, 2011, Delhi official). Thus, the exit of this community from the Valley is understood as rooted in a simple family choice, driven by family's own calculus. Consequently, the crisis of KP displacement came to be labeled as "migration" and those displaced as the "migrants." These positions convey a nuanced perspective of displacement as such labeling has policy implications by impinging on the rights of protection and return, extended to those who are traditionally recognized as having become "internally displaced within their own countries." The framing of this displacement as a migration has led to the creation of piecemeal or stand-alone policies, suitable for myriad migratory situations, such as in the case of economic migration, in contrast to the holistic and long-term policies for those ousted from conflict-prone societies. Such positioning has also liberated the officials from confronting the perpetrators and addressing the malfunctioning of the state-level institutions, thus leaving the societal issues unaddressed and unresolved.

Consequently, as dictated by the top-down nature of societies, these practices have placed the KP families on the bottom rung of the society (Galtung, 1969), sending them into deeper marginalization. However, the authorities tout the gains secured by the families as a result of living in the metro communities. The most significant aspect of the migrant positioning of the families has been the caveat that "those who voluntarily migrate cannot be made to return against their own will" (personal communication, August 1, 2011, Delhi official). The leaders suggest that asking the families to return through official policies would be against the wishes of the families and a violation of the future they had envisioned when they chose to leave the Valley. The consequent provisional policies are justified as the best humane response to the unanticipated migration of large number of families. However, the families feel humiliated and object to the labeling of their community as migrants, realizing the deeper implications of such labeling as a denial of their right to return to ancestral communities. Additionally, the migration explanation of the crisis has branded the families as poor and helpless, thwarting their efforts to improve their standing in host communities.

As exemplified here, the conflicting positions embraced by the leaders to articulate the KP displacement has significantly influenced the outcome of policies, resulting in policies ill-suited to address the issues of the families and an absence of policies to address the root cause of the community's ouster.

Civil society positions

The national leaders are not alone in maintaining their steadfast and yet conflicting positions on the KP displacement, which have come to influence the outcome of policies. Given the long-term duration of this crisis, members of the civil society and the family advocacy groups have also been challenged on how to accurately articulate the crisis and represent the evolving needs of the families to ensure the rollout of meaningful policies. Immediately after the crisis, there was a proliferation of grassroots organizations, which emerged to smooth out myriad administrative and operational hurdles required to respond to the immediate needs of the families. These family advocates appeared in many forms and sizes, facilitating the family registration process, setting up camps, assisting the families to navigate new societies, and securing identity cards for the families.

During the early phases of displacement, the officials as well as the families credited the work of these groups as instrumental in successfully securing important concessions for the KP families, such as the securing of suitable accommodations, facilitating the provision of medical benefits, and identifying suitable schooling for the KP children. Subsequently, the authorities came to rely on these organizations in the crafting of family-oriented policies. However, as the fatigue from extended displacement began to take over, several of these organizations folded or came to take up different tasks per the shift in their own organizational missions and mandates. Whereas the initial positions advanced by these organizations were understood to have resulted in beneficial policies for the families, the shifting organizational priorities came to be perceived by the families as resulting in premature and useless policies that were misrepresenting the families' genuine needs. Rather than being at the helm of the KP families as their savior, these groups began to be perceived as threatening the KP community's unity.

Family positions

The family positions that come to impact the outcome of policies have been largely influenced by the family's perception of the underlying motives of the lawmakers as well as their own advocates and members of the civil society. During their prolonged displacement, the families have become suspicious of the motives of their own supporters, perceiving them as weak and having remained outside of the sphere of policy. Positioning themselves as victims and as powerless, the families view their advocates through the same lens, as those "with limited voice, and outside of the official dialogue" (personal communication, July 27, 2011, Delhi-based family). The families' negative perception of their own advocates is also driven by the absence of a robust civil society in the Valley that would have prevented their ouster in the first place. Dr. Sazawal (2009), an authority on the predicament of the displaced KP community, explains that

> whereas civil societies, around the world, gain prominence and standing through their contributions to societal and human development, by creating space for diversity, instead, the one-dimensional culture of Kashmir's civil society, focused on politics, may have destroyed the KP families' aspirations for a balanced and wholesome development of their society.

Some families see the advocacy groups as intentionally misrepresenting the family voices and attribute the government's lack of a "long term vision for the families to the failure of these groups" (personal communication, July 23, 2011, Jammu township). Collectively, the plethora of advocacy groups are said to lack a unified voice to represent the families and are said to be operating with bias and their own political agenda. For instance, one such group that claims to be speaking on behalf of the majority of the KP families has advocated for a separate homeland, stating this to be the only legitimate solution for the families from the security perspective. Simultaneously, a separate group has continued to push for the return of the families to their ancestral homes, declaring this to be the wish of the majority of the families and the only dignified solution which "honors the families undeniable right of return" (Rajput, 2012). The mixed signals from the advocacy groups have often stifled the initiatives of national actors, who face their own dilemma of whether to modernize and invest in new settlements or work with the Valley administrators to secure the families' safe return. Further, the families claim that the conflicting agenda of their advocates often creates rifts not only within their larger group but also within the families, declaring the work of these organizations to be "harmful" and unsuitable to addressing the unique circumstances of the KP families. This has resulted in the family's own inability to push for stronger and more effective policies as a cohesive community with one voice.

Conclusion

This chapter has exposed subtle clues behind the positions embraced by key actors of displacement that come to influence the effectiveness of IDP policies. It has become clear that the positions that are embedded in the interests of the actors themselves often result in unintended consequences, thus negating the well-intended solutions and policy outcomes. The mixed signals sent through the conflicting narratives also undermine the genuine effort of the actors. From the KP family perspective, a host of policies has neither calmed the anxieties of those who are understood to have voluntarily migrated nor pacified those who are understood to have escaped a temporary instability. The multiple positions advanced by their advocates have also contributed minimally to the psychosocial well-being of the KP families.

As can be detected from this chapter, IDP policymaking in general and the KP policymaking specifically, given the prolonged nature of displacement, remains an enduring challenge. Extended displacements come to be constrained by a host of variables, such as the changing needs of the families themselves, resistance from the family advocates, resistance from the host communities, confrontation from those who never left their hometowns, and the country's politics embedded in the country's national agenda. The following chapter takes up the issue of the complexity of IDP policymaking, as complicated by these and numerous other variables.

References

Galtung, J. (1969). Violence, peace, and peace research. *Journal of Peace Research*, 6(3), 67–191.

Harré, R., & van Langenhove, L. (1999). *Positioning theory: Moral contexts of intentional action*. Malden, MA: Blackwell.

Rajput, S. (2012). *The displacement of the Kashmiri Pandits: Dynamics of policies and perspectives of policymakers, host communities and the Internally Displaced Persons* (Doctoral dissertation). Fairfax, VA: George Mason University. ISBN: 9781267843333.

Sazawal, V. (2009, April 30). *The role of civil society in Kashmir*. Retrieved from Kashmir Forum.org

Wistrand, J. (2013). *The role of Azerbaijan's post-conflict national narrative in limiting refugees' and IDPs' integration into mainstream society*. Retrieved from Webcast www.wilsoncenter.org

9 Complexity of IDP policymaking

Introduction

This chapter takes the readers behind the policies and into the mindset of the policymakers. A thoughtful understanding of the myriad roadblocks encountered in the formulation of displacement policies not only delivers a deeper and a much-required understanding of the predicament of those who endure displacement but also brings to bear a unique appreciation of the national policymaking process itself. The bottlenecks encountered by the lawmakers in the formulation of national policies suggest a context-specific and a thoughtful engagement of international actors in situations of mismanaged, extended, or unresolved situations of internal displacement.

Policymaking for communities of the internally displaced persons (IDPs) in general and the displacement of the Kashmiri Pandits (KPs) in particular remains a challenge, constrained by a host of variables. Given that the phenomenon of internal displacement and specifically conflict-induced displacement is generally understood to be an outcome of a society's own mismanagement of its people, and, consequently, the handling of those displaced within the society, a responsibility of the local leaders, a clear understanding of IDP policymaking remains obscure.

This chapter exposes the hidden obstacles into the formulation of a nation's displacement policies and specifically the obstacles to the development of KP policies. These bottlenecks can be attributed to a host of factors, such as an absence of a universally and consistently applied framework; the growing as well as the changing needs of the families; resistance from family-supported organizations; resistance from the host communities; protests of equity from those who never left the hometowns' communities, even at the height of mass violence and turbulence; and, most importantly, the limitations posed by the country's own national positions, agenda, and goals, and the country's own understanding of the concept of internal displacement.

The following paragraphs help illustrate the nature and the rationale behind these obstacles, thus advancing an understanding of how these roadblocks come to impact the formulation and rollout as well as the outcome of IDP policies.

Absence of universally applied framework

As eluded to in earlier chapters, the complicated formulation of IDP policies contrasts sharply with the universally upheld international framework in place for those who seek protection outside of home countries and come to be classified as "refugees." The policy framework for the latter group is guided by the *1951 Refugee Convention* and the subsequent *1967 Protocol Relating to the Status of Refugees* (UNHCR, n.d.), a legal international document ratified by over 100 sovereign nations. This document clearly identifies the legal obligations of member states to protect the refugees. The protection provided under this Convention is "universal in scope with the responsibilities of the supporting governments equitably distributed and consistently applied across nations" (pp. 1–5). The United Nations High Commissioner for Refugees (UNHCR) serves as the "guardian" of the Convention, ensuring that sovereign states cooperate in safeguarding the rights of refugees. The Convention is grounded in humanitarian principles of protection and assistance for the refugees. Specifically, it guarantees the movement of people, facilitates issues of their resettlement, and ensures family unity and provision of welfare services for all those under the United Nations (UN) protection system, regardless of the reasons for their displacement (UNHCR, 1951, Convention relating to the Status of Refugees).

However, in the context of those who remain internally displaced within the borders of their own countries, there is an absence of such legally binding mandate, except as contained in the *Guiding Principles on Internal Displacement* launched in 1998 (UNHCR, 2004). In contrast to serving as a universal mandate supported by enforced mechanisms and ratified by member states, these Guidelines serve as a tool for "raising awareness of the needs of IDPs, mobilizing support within the humanitarian community and assisting individual governments in providing for the security and wellbeing of their displaced populations" (FMR, 2008; OCHA, 2015). Consequently, these Principles only serve as suggestive guidelines for "the most vulnerable of the human family" (OCHA, 2015). As explained in the following paragraphs, an absence of a universally binding framework introduces several sources of resistance, stemming from different constituencies, that come to stifle the formation as well as the effectiveness of national-level displacement policies.

Obstacles to KP policymaking

Several obstacles to the formulation of displacement policies for the KP families have rendered the process of policymaking complex and obscure, often resulting in a set of ad hoc and ambiguous policies. As unveiled in earlier chapters, the KP policy framework remains largely guided by the initial storyline embraced by the officials to explain the unprecedented exodus of this community. It has also become clear that over time, the random policies that have evolved have not entirely addressed or pacified the concerns of those who have remained displaced for nearly three decades. However, the gridlocks to the KP policy formulation cannot be entirely attributed to the positions embraced by the lawmakers or the absence of universally enforced rules or mandates. Instead, over the life cycle of this displacement, the obstacles to policies have stemmed from numerous additional sources. Hindrances to the KP policies stem from factors such as the naturally evolving and growing needs of the families themselves; resistance from KP-supported groups; and resistance from members of host communities, whose voices heavily factor into the design as well as the implementation of key displacement policies. In several instances, host families for the KP communities have been able to strongly influence key IDP policies that have affected the lives of displaced families at some point during their extended exile, such as the policies relating to housing accommodations, education, and jobs.

In addition, the policymakers have faced resistance from those who remained in the Valley as well as the perpetrators of the families who ousted the KP community. The families' own positions also cannot be ruled out as having stifled the policies as well as the policymaking process. Given the lapse of time, the KP families themselves have become conflicted as to what may constitute a helpful and beneficial policy, given their own changing circumstances and evolving vision for the future. The array of these disorders to the KP policy formulation are worth examining as they provide direct clues into the context-based obstacles to policymaking and may indirectly point to the context-specific solutions.

National constraints

Policymaking for the KP community has been tricky and challenging since the onset of this exodus and the families' first arrival into new communities. The preliminary hurdles faced by the families can be attributed to the limitations of the administrative systems, which were found to be ill-equipped to cope with the unprecedented and the unexpected exodus of this community. The hurdles faced by the camp administrators in setting up the family registration and designing systems for identification and the ration cards contributed to the initial policy hurdles. However, the subsequent evolution of a host of ad hoc policies emerged from the recognition of this displacement as an event of a short-term duration, resulting in a collection of transitional policies. Over the decades, this understanding of the KP exodus prevented the policymakers from securing more robust solutions for the families. The full rehabilitation of the families into their host communities would have meant a dramatic shift in the admission of the crisis from one of a temporary nature to one that may have completely damaged the home community and may thus be of an irreversible nature.

Additionally, an undertaking toward more durable solutions would require an acknowledgment of the failure of official negotiations with the KP community's perpetrators in the Valley responsible for ousting this community and an admission that the home communities remain unsafe for return. More poignantly, the emergence of durable policies would render the return and rehabilitation policies that have been in the making irrelevant and subject to perception by the families as a fabrication, an insult, and a ploy to pacify those longing to return. The combination of positions, the optics, and the lawmakers' fatigue with the protracted displacement have rendered the KP policymaking a grueling task for the officials.

Family resistance

As the size of the families grew over the extended displacement, the families' needs for larger accommodations, more suitable schools for the growing children, and assurances for job security became the families' new concerns. Resistance to new policies often came from the families themselves, such as the KP Delhi-based families, who resisted to the location of the new housing in *Dwarka* on the outskirts of mainstream Delhi. The families objected to the location, suspecting it to be "solely driven by the pressure of the host communities' intention to keep the KP families out of mainstream and thus away from civilization" (personal communication, July 27, 2011, resident of Dwarka camp). The families also voiced their suspicion of the return policies that offered job incentives for those willing to return to the Valley, suspecting these policies to be "politically motivated and insincere" (personal communication, July 27, 2011, resident of Dwarka camp). Consequently, a comprehensive return and rehabilitation policy never came to fruition and attempts to address the return of the families remains dismissed as ingenuous and never fully embraced nor given a chance for refinement or debate that could, at a minimum, offer a platform to weigh the strengths and gaps of this policy.

Host community resistance

Lawmakers also faced several hurdles in upholding their commitment to the KP families, given the resistance posed by members of the host communities. In the early years of displacement, the families were given accommodations in Delhi's *Bappu Dham* community, in the northeastern district of Delhi. Suspecting that the KP community's stay could last for longer period, the locals of the community banded together and petitioned their local lawmakers to have the families vacate the community halls (*Barat Ghar*) that were being used to house the KP families. This left the policymakers scrambling to explore alternative solutions for the families. In other cases, the locals often opposed the building of the "migrant schools" in their communities, objecting that the building of schools for the KP children would create the appearance of a migrant and thus a poor community. The official plan to provide some means of livelihood for the KP families, such as through the use of the temporary shops in select host communities, also did not come without resistance from the business community of host societies. Delhi's area businesses, in the city's INA market (Indian National Army Market), protested to the free use of the prime real estate by migrant families (Rajput, 2012) that were supposedly only going to be in Delhi for a temporary stay.

Hometown community resistance

Not only did the policymakers face influential resistance from members of the host communities, but equally forceful was the resistance from members of the KP's own hometown communities. Strong opposition to new policies often came from those who had remained in the Valley, who saw their residual community as the one that was truly marginalized and in need of development. Those who had fled were understood to have secured better opportunities in metropolitan communities by default. The hometown communities protested on occasions, when KP housing or education policies were unveiled by the government, objecting to those policies as "unfair deals" for those who continued to live in fear of being ousted. Resistance to official policies also came from the perpetrators who had ousted the

KP community. This group denounced any KP policy through street demonstrations and burning of flags, creating a threatening environment for the hometown residents. As a result, the policymaking as well as the rollout of any policies became a secretive undertaking, often leading to untimely, jumbled, and vague policies that the families as well as the implementers had a difficult time deciphering.

Civil society resistance

Although the national government remains the official policymaking body for KP policies, several policies have been informed by the family-supported groups. The family advocacy groups have been credited with having secured several important concessions for the families in the early part of their displacement. However, as displacement prolonged and with the proliferation of the advocacy groups, different groups came to represent different sets of needs and wishes of the families. In the absence of a unified voice that represented the KP community as a whole, some groups objected to the rollout of certain policies as unfair and useless, whereas other groups accepted and embraced the same policies. The disconnect within these groups has been most pronounced in the context of return and rehabilitation policies. For instance, some local non-governmental organizations (local NGOs) have advocated for an immediate return of the families to their ancestral homes, thus stifling the government's initiative of building a dedicated community for the families until they are ready to return. Other KP-supported groups have demanded a separate homeland for the returning families, thus discouraging any attempts by the officials to reconcile with the families' perpetrators. The officials are of the opinion that the positions maintained by the family supporters have stifled the official strategy, thus derailing the uplifting of the plight of the KP families, while they remain outside of home communities.

Conclusion

As unfolded here, the KP policymaking, and in general the policymaking to address violence-induced displacements, remains a challenge, largely owing to the fact that the management of internal displacement is assigned to the national actors. While striving to secure the best possible solutions for the displaced families within their sovereign territories, the national leaders find themselves constrained by a host of internal bottlenecks and resistance that flow from all levels of the society. Such tribulations make the task of IDP policymaking a sensitive and complex undertaking, which signals an appropriate, context-specific, and thoughtful need for international guidance.

While respecting and recognizing that the challenges and issues unfolded in the preceding chapters are unique to the culture and the nature of the hardships that the KP families encounter, in order to gain a deeper appreciation of the predicament of the KP families, issues of similarly displaced communities across the globe are taken up in the following Section. Not only does this effort help situate the displacement of the KP ethnic community in the context of similarly displaced communities, but also, more importantly, this indulgence represents a unique effort in displacement literature to zero in on the core IDP experience. The tracing of commonalities and individualities across communities offers possible context-driven clues into societal and policy reforms and solutions.

References

Forced Migration Review. (2008, December). *Ten years of the guiding principles on internal displacement.* Refugee Studies Center, Oxford, England.

OCHA. (2015, January 22). *The forgotten millions.* United Nations Office for the Coordination of Human Affairs.

Rajput, S. (2012). *The displacement of the Kashmiri Pandits: Dynamics of policies and perspectives of policymakers, host communities and the internally displaced persons* (Doctoral dissertation). Fairfax, VA: George Mason University. ISBN: 9781267843333.

UNHCR. (2004). *Guiding principles on internal displacement.* (OCHA/IDP/2004/01).

UNHCR. (n.d.). 1951 Convention and protocol relating to the status of refugees.

Section IV
Understanding Kashmiri Pandit displacement through a comparative lens
Journeying into Azerbaijan, Georgia, Serbia, and Sudan

Introduction to comparative cases

In an effort to situate the displacement of the Kashmiri Pandits (KPs) in a comparative perspective, this chapter launches a four-chapter exploration of the similarly displaced communities of the Nagorno-Karabakh region of Azerbaijan, and the Abkhazia and South Ossetia regions of Georgia; the displaced community of ethnic Serbs from Kosovo; and the families displaced from Sudan's Darfur region. Although spread across the global landscape, these four distinct communities bear resemblance to the displacement of the KP community. Similar to the KP displacement, the displacement of these communities was triggered by conflict-induced elements as opposed to natural disasters; those displaced were forcefully evicted by members of their own societies; the magnitude of displacements is comparable; and, most glaringly, time in exile for members of these communities is also of a protracted nature. The analytical framework that has helped unfold the interconnectivity of the KP family challenges (Rajput, 2012), shedding light on the political and the community entrapment of the families, namely, Dugan's Nested Model (Dugan, 1996), will be leveraged once again to systematically decode the challenges of these communities.

The journeying into these communities begins by demystifying the challenges of those displaced in Azerbaijan (current chapter), where close to half a million people remain displaced after more than two decades, displaced from homes in the Nagorno-Karabakh region, now a self-proclaimed republic. Next, the journey continues with the discussion of those who have remained displaced for over two decades within Georgia (Chapter 11), ousted from their ancestral homes in Abkhazia and South Ossetia, the areas that became the target of secessionist influence in the early 1990s. Next, the challenges of those who remain displaced in Serbia (Chapter 12), having fled the ethnic tensions in Kosovo in 1999, are unveiled. This global journey concludes with a thoughtful examination of the plight of the families from Sudan's Darfur area (Chapter 13), from which the violence that erupted in 2003 displaced more than three million people.

10 Azerbaijan
Displaced from Nagorno-Karabakh

Introduction

According to Azerbaijan's State Committee for Refugees and Internally Displaced Persons (IDPs), about half a million people remain displaced after more than two decades, ousted from the country's Nagorno-Karabakh region (IDMC, 2014). Azerbaijan is a sovereign state bounded by the Caspian Sea and Caucasus Mountains, which span the continents of Asia and Europe. The country was one of the first republics of the former Soviet Union that faced the problem of internal displacement. The country currently has one of the highest numbers of IDPs per capita in the world, with 7% of its population in internal displacement (UNHCR, 2009), residing in all 76 districts of the country.

Capitalizing on the diagnostic tool kit used to decode the challenges of the displaced KP families (Rajput, 2012), the Nested Model (Dugan, 1996) will be used once again to unravel the challenges of the displaced Azeri families and, uniquely, for the first time, to assess the KP displacement vis-à-vis displacements of those displaced from the Nagorno-Karabakh region. The following sections delve into the circumstances of eviction of this minority community; the challenges encountered by the families; key policies and their perception; and the dilemma of return that keeps the families in a state of perplexity, similar to the anxiety faced by the KP families.

Circumstances of displacement

As the nationalist aspirations among the people resurfaced in 1988, civil unrest took hold of the Nagorno-Karabakh region. The Soviet government of the neighboring country of Armenia declared that the Nagorno-Karabakh territory was to be incorporated into Armenia. The conflict over the origin of the region between the two countries

92 *Understanding Kashmiri Pandit displacement*

triggered the displacement of about half a million ethnic Azerbaijanis (IDMC, 2014). The ancestral roots of those who fled remain in the area now controlled by the self-proclaimed Nagorno-Karabakh Republic, while living in exile for over 26 years, similar to those displaced from the Kashmir Valley. The circumstances of eviction continue to impact the families' physical, mental, and spiritual domains as many have been impacted by health and disability issues, some suffer from the psychological trauma of forced eviction, and many continue to find it difficult to live with a shaken faith and the diminishing prospects of their return.

Challenges encountered

The analysis and the interpretation of a host of challenges that the Azeri families continue to encounter, as framed around the Nested Model, are presented in the following paragraphs.

Specific issue

As explained in earlier chapters, the *specific* issue of the comprehensive conflict system in the framework of the Nested Model addresses the immediate issue faced by the people (as illustrated in Figure 5.1). Similar to those displaced from the Valley, the *specific* issue of the displaced Azeris centered on their having lost their ancestral homes in the Nagorno-Karabakh region. Initially, most of those displaced lived in tents and public buildings such as schools, hostels, and dormitories (UNHCR, 2009). After 26 years, some families continue to live in unsafe shelters, established from railway wagons, basements, and areas below football stadiums, put together with scrap materials and without access to utilities or sanitation (IDMC, 2014). The families' sense of identity now includes an aspect of their having lost their homeland (UNHCR, 2009); consequently, the experience of having become homeless now pervades their inner being and has become the *specific* issue of their displacement in the context of the Nested Model.

Relational issues

In the context of the Nested Model, *relational* issues refer to the *fit* of the people into a community. Numerous bottlenecks have thwarted the families' adjustment as well as integration into new host communities within Azerbaijan. Initially, as families relocated to new communities, many were forced to split as the male members of the families left in search of jobs elsewhere in the country (IDMC, 2014). In addition, the school segregation policy reinforced the families' stigma and kept them isolated from the broader community, hampering their integration and adding to their precarious social position (Wistrand, 2013). The elderly population has been particularly affected, as the prolonged displacement has led to their isolation, eroding their psychological and social well-being. The families' participation in public life has also been limited by the policies which restrict their voices in municipal matters, such as in the communities' local elections.

The officials and the administrators justify these segregated policies on the ground that the families are eventually expected to return to their hometown communities, where the families will regain all their previous rights. Additionally, the officials maintain that the segregation policies are aligned with the families' own preferences to preserve the cohesion of their community (IDMC, 2014). However, an absence of voice on issues of local importance has limited their access to local resources and has stifled the Host/IDP relationship. Although, over the years, relations between IDPs and the locals have become more amicable, with the locals having become more "tolerant of the families" (UNHCR, 2009), tensions between the two groups continue to surface, given the host families' perception of the displaced families as enjoying superior privileges. These privileges refer to access to employment, government assistance, and free health services, which have helped reinforce the belief among the locals that the *Presidential Decrees* have afforded the IDPs better access to certain rights than afforded to the locals. This sense of "relative deprivation" (Gurr, 1970), where one group perceives the other group to be faring much better in spite of their own superior capabilities, is similar to the sentiments expressed by members of the host communities of the KP families. The host communities of Delhi and Jammu had also come to perceive the KP families as being afforded a number of free services, such as housing, education, and job quotas, services for which the host community had to pay and work for.

Subsystem issues

Subsystem issues within Dugan's Model refer to those issues that lie underneath the surface of the conflict but come to influence the conflict dynamic, such as the varying value systems or the education systems of the Host and the IDP communities. In cases where there are significant differences, these variances come to negatively impact the adjustment of the displaced families. Most of those displaced from the Nagorno-Karabakh region came to new communities with an agricultural background and a lower level of education and less developed skills (UNHCR, 2009) as compared to the locals of their host communities. These inequalities delayed the families' adjustment into their new communities. Given these differences, the displaced youth found it hard to link their academic and professional qualifications to the demands of the new communities. Further, during the years of protracted displacement, their previously acquired skills weakened, making it hard to either locate jobs or to continue with their education.

Structural issues

Structural issues of displacement refer to the overarching mechanisms and systems that govern the families' access to basic services, which can either promote the well-being of the families or push them into further marginalization (Galtung, 1969). In an effort to manage urban migration and following the legacy of the Soviet "propiska" system (UNHCR, 2009), the government policies have restricted the families' migration to rural areas. The remote location of the settlements has also made it difficult for the displaced Azeri families to access essential services, such as the medical facilities or the schools for their children. The families that change their registered residence also fear being kicked out of the assistance programs. This has made it difficult for the family as a unit to move to urban areas in search of better opportunities, keeping them in marginalization (Galtung, 1969). In addition, given the family's temporary residence status, the families face difficulties in obtaining rights of residential property ownership. Consequently, the families in rural areas remain dependent on government assistance, which has taken away the sense of self-esteem and worth that comes from a feeling of self-reliance (Korostelina, 2007).

The families' challenges, complicated by the location of the settlements, access to schools, or restriction in the movement of the families, are grounded in the labeling of the families as *temporary* residents, mirroring the official narratives that have been used to label the KP families. These practices have kept the families on the bottom rung of the society, as dictated by the top-down vertical nature of the society's governance (Galtung, 1969).

Relevant policies

The Azeri government is increasingly assuming greater responsibility for the care and protection of the displaced families, and has been devoting significant resources. Official efforts have included raising awareness of the displaced families, training officials on IDP rights, adopting laws, and instituting several protection programs (IDMC, 2014). In 1993, the *State Committee for Refugees and Internally Displaced Persons* was also established. The legal framework includes the 1999 Law on *Social Protection of Forcibly Displaced Persons* and the 2004 State Program for the *Improvement of Living Standards and Generation of Employment for Refugees and IDPs*. In 2013, the country's State Oil Fund also earmarked about US$600 million to improve IDPs' living and housing conditions, and over the past two decades, the authorities have spent US$5.4 billion for the families' needs (World Bank, 2013). The impact of specific policies is detailed in the following paragraphs.

Housing

Similar to the official positioning of the KP community's exodus, the "temporary" designation of the Azeri displacement has allowed for the provision of cost-free "temporary" housing, while their permanent residence is recognized as their hometown community. A number of families continue to live in the former administrative buildings and the collective centers that were offered as temporary accommodations at the outset of displacement more than a decade ago. However, over the past decade, the government has dedicated significant resources to new housing, transferring some families to housing with schools and medical centers (IDMC, 2014). The 1999 Law on Social Protection of IDPs stipulates that IDPs are to be provided housing with the "temporary" nature of residence. In the spirit of family unity, relatives of families have also been relocated from several regions and have been allowed the use of nearby land for livelihood activities. The plan is to provide new housing to around 70% of the country's IDPs. However, the new housing is allocated on a temporary basis until the families are able to return, restricting the families to rent or sell the property. In an effort to maintain social cohesion among the displaced families, the new housing also remains isolated from the host communities.

Education

Education policies are also embedded in the official narrative that the families are eventually to return to their hometowns. The children are offered cost-free education separately from the children of the mainstream community (Radsky, 2013). The government's justification for the segregated school system is to help the children to adapt to their displacement and to maintain the social fabric of displaced communities, which will facilitate their integration on eventual return to their hometown communities.

Jobs

State benefits, such as the monthly allowance and social assistance programs, remain the main source of income for more than 70% of Azeri families (World Bank, 2011). Those who were public sector employees at the time of their displacement have been allowed to keep their posts, while others receive monthly compensation. A large number of families have benefitted from the employment quotas as well as low-interest credits (ICG, 2012). However, the unemployment rate for IDPs remains high compared with the national unemployment rate as families remain reluctant to move to other towns in order to find jobs, given that their monthly allowance and housing is linked to their continued presence at the government-registered address (UNHCR, 2009).

Perception of policies

For the displaced Azeri families, the rural and remotely located settlements continue to enforce their sense of impermanence, discouraging their embrace of the new community or their willingness to integrate. The isolated community stigmatizes the families as poor and helpless, obstructing their efforts to improve their self-reliance (Wistrand, 2013). The government's rationale for linking the assistance to IDPs' registered addresses is intended to facilitate the families' reintegration upon return; however, the families perceive this as a roadblock to their growth opportunities. The families have likely benefitted the most from the tuition-free education extended to their children. They are also of the opinion that the segregated schools do not reflect discriminatory policies and are not problematic; in fact, the families are appreciative of the segregated schooling as it has allowed them to retain their status and maintain their cultural heritage. The segregated schooling has also provided the families with employment in the schools. The families perceive the education policy as instrumental in eventually lifting the families out of poverty (UNHCR, 2009).

Issue of return

The issue of return remains the most problematic of the displacement challenges for the Azeri families. The return policies remain embedded in the political *positioning* of this displacement as temporary, with policies revolving around the idea of a possible return. In 2005, the government drafted the principles of return and coordination mechanism in the event of a peace deal and set aside large sums of money for the return of half a million people (Johansson, 2010). Despite the fact that, in the absence of other options, several families have integrated to a certain extent, these arrangements do not represent durable solutions from the official perspective as doing so would signal abandoning the claim to the disputed territory (Johansson, 2010, p. 71). Although the two neighboring countries, Azerbaijan and Armenia, have maintained relative peace, the absence of a permanent resolution leaves the fate of this ethnic minority in limbo. The elderly long to return home, if only "to be free to die in the land they cherish and miss so much" (Matiash, 2014). Residents of public buildings find little value in renovating their housing, given their understanding that their situation is temporary and that they do not own the property. The young and the elderly alike see these settlements as a "temporary situation before they can finally return and rebuild their lives in their own lands" (Matiash, 2014). The majority of IDPs themselves wish to return, but given that the prospect of a peace settlement remains remote, and the families have endured their current conditions for over two decades, they consider their situation to be far from temporary.

Azeri families juxtaposed with Kashmiri Pandit families

A comparison of the Azeri community's challenges with those displaced from the Valley, the KPs, unveils greater similarities than contrasts, as explained in the following section.

The most troubling of the similarities is the reality that after more than two decades of displacement, some segments of both ethnic minorities remain displaced, living in extended internal exile within their own countries. The *specific* issue for both communities, as the families were moving out of their ancestral homes for the first time, centered on their having lost their family homes, thus creating the phenomenon of homelessness.

Similar to the official positioning of the KP exodus as an event of a temporary nature, "temporary disturbance" displacement policies for the Azeris also revolve around the idea of a possible return to their permanent homes. Accordingly, the families are provided "temporary" housing, with their place of origin in the Nagorno-Karabakh region considered their permanent residence. Any new housing is allocated on a temporary basis until the families are able to return, restricting the families to rent or sell the property. This housing design resembles the "Migrant Township" model for the KP families, where the government retains the ownership and control of the housing complex. Consequently, the labeling of the families in both communities as temporary residents has exaggerated their overall adjustment challenges, keeping them on the bottom rung of their host societies. Azeri families' perception of policies ranging from being stigmatized and keeping them isolated to being helpful in their eventual escape from poverty is similar to the assessment of the policies by the KP families, which ranges from a "blessing in disguise" to "total humiliation to the very being." The adjustment and the coping skills of both communities have been hampered by the circumstances of their conflict-induced eviction. The experience of having been forcibly evicted and the recall of the day of departure now strain both communities' psyche.

However, adjustment into their host communities was more trying for the Azeri families as most entered their new communities with a farming background and a lower level of schooling and less developed skills (UNHCR, 2009) in comparison to the locals of their communities. In contrast, the displaced Kashmiri families brought a spectrum of skills to their new communities, having served as teachers, doctors, writers, and government employees in the Valley before displacement. The Azeri youth also found it difficult to quickly become relevant to the demands of their new communities (UNHCR, 2009). In contrast, the government-sponsored educational opportunities afforded to the Kashmiri families, through the *Special Allocations for the Migrants*, allowed the KP youth to resume their interrupted education quickly.

Nevertheless, the senior population remains equally affected in both communities as the segregated housing policy, justified on the basis of temporary stay, has limited their interaction with the wider communities, leading to their isolation and boredom. The Host/IDP dynamics have been contentious for both communities, given the host communities' perception of the families as enjoying benefits, extended through the *Presidential Decrees* in the case of Azeri families and the *Special Allocation for Migrant Families* for the KP families. The overarching mechanisms that govern families' access to basic services have also been limited for both communities.

The issue of return remains the most problematic and enduring of the displacement challenges for both communities. The return policies remain embedded in political posturing of the two displacements as of momentary nature. Although the ancestral roots of these communities reside in the area from which they fled, the return to those communities remains more problematic for the Azeri families as the area of their former residence has now been informally incorporated into a new territory, now controlled by a self-proclaimed republic. However, even with the diminishing prospects of return, the families in both communities cling to the hope of eventual return.

Conclusion

After more than two decades, the displaced Azeri as well as the KP families lack policies that support their full integration, as the current policies are aligned with the official positioning of the two displacements as "temporary," which has ruled out the families' local integration in the host communities (IDMC, 2014). The government's stance on return has hindered the restoring of IDPs' rights and improving their self-reliance and integration while they wait to return. In the meantime, both communities remain ambivalent, feeling a complete loss of identity and yet clinging to an identity that keeps them hopeful of returning to the very communities that ousted them decades ago.

The following chapter unfolds a similar predicament of those who remain displaced after more than 26 years, forcibly evicted from homes in Georgia's South Ossetia and Abkhazia regions, the areas that became the target of secessionist influence in the early 1990s.

References

Dugan, M. (1996, Summer). A nested theory of conflict. *Women in Leadership, 1*(1).

Galtung, J. (1969). Violence, peace, and peace research. *Journal of Peace Research, 6*(3), 167–191.

Gurr, T. (1970). *Why men rebel*. Princeton, NJ: Princeton University Press.

ICG. (2012, February). International Crisis Group 'Tracking Azerbaijan's IDP Burden'. Briefing No. 67. Retrieved from https://www.crisisgroup.org/europe-central-asia/caucasus/azerbaijan/tackling-azerbaijan-s-idp-burden

IDMC. (2014). *Azerbaijan: After more than 20 years, IDPs still urgently need policies to support full integration*. Geneva, Switzerland: Internal Displacement Monitoring Center.

Johansson, P. (2010). *Peace by repatriation concepts, cases, and conditions*. Umeå: Umea University.

Korostelina, K. (2007). *Social identity and conflict: Structures, dynamics, and implications*. New York, NY: Palgrave Macmillan.

Matiash, C. (2014, August 15). *Searching for home: Ed Kashi's Photos of IDPs in Azerbaijan*. Retrieved from www.blogs.wsj.com

Radsky, V. (2013). *Developing inclusive social policies: Education for Azerbaijan's Internally Displaced*. Baku, Azerbaijan: Center for Innovations in Education.

Rajput, S. (2012). *The displacement of the Kashmiri Pandits: Dynamics of policies and perspectives of policymakers, host communities and the Internally Displaced Persons* (Doctoral dissertation). Fairfax, VA: George Mason University. ISBN: 9781267843333.

UNHCR. (2009, October). *Azerbaijan: Analysis of Gaps in the Protection of Internally Displaced Persons (IDPs)*. Retrieved from unhcr.org

Wistrand, J. (2013). *The role of Azerbaijan's post-conflict national narrative in limiting refugees' and IDPs' integration into mainstream society.* Retrieved from Webcast www.wilsoncenter.org

World Bank. (2011). *Azerbaijan: Building assets and promoting self-reliance: The livelihoods of Internally Displaced Persons* (Report No. AAA64–AZ.)

World Bank. (2013). *Implementation status & results Azerbaijan IDP living standards and livelihood project* (Report No. ISR9036).

11 Georgia
Displaced from Abkhazia and South Ossetia

Introduction

Building on the last chapter, this chapter offers an additional opportunity to deepen an understanding of the challenges of the displaced Kashmiri Pandit (KP) community through a comparative perspective. This chapter delivers a systematic and a comprehensive understanding of the displacement of the ethnic communities that fled the towns of Abkhazia and South Ossetia within the Republic of Georgia. At the outset, the displacement of the Abkhazian and the South Ossetian communities displays critical similarities to the ouster of the KP community, specifically the triggers of displacement, the magnitude of those who fled, and their time in internal exile, all amounting to a comparable level of human suffering by both communities. These features lend legitimacy to the discussion of this case for the purpose of comparing the two communities.

The Republic of Georgia, a former Soviet Republic, with a population of 5.4 million people, is situated at the crossroads between Europe and Asia, and has had its share of the internally displaced persons (IDPs) as a result of multiple conflicts. Similar to the displaced KPs and the Azeris of the Nagorno-Karabakh region, more than a quarter million people, (Rajput, personal communication, May 13, 2015, Ministry for Internally Displaced Persons from the Occupied Territories, Accommodation and Refugees of Georgia) had been ousted from their ancestral homes in Abkhazia and South Ossetia, the areas that became the target of secessionist influence in the early 1990s. For over two decades, these Georgian families have remained in long-term displacement, dispersed throughout the country. The analytical framework, rooted in Dugan's Nested Model (Dugan, 1996) that has successfully unveiled the overlapping dimensions of the displacements of the KP minority and the Nagorno-Karabakh communities, will be put to test once again to help decode the challenges of the Georgian families.

Circumstances of displacement

In the early 1990s, the call for secession from Georgia in the two autonomous regions of Abkhazia and South Ossetia, from the country which pursued to preserve its territorial integrity (IDMC, 2012), triggered the displacement of ethnic Georgians from their ancestral homes. Subsequent conflict in 2008 left the de facto authorities in the two regions in complete control of these territories, supported by the Russian Federation, in their calls for independence. The government of Georgia considers South Ossetia and Abkhazia as its own territories that are currently being occupied by Russia. These armed conflicts remain unresolved, and as of end of 2014, more than 232,000 people remained displaced (IDMC, 2012).

Challenges encountered

Having been forcefully evicted from conflict-induced circumstances, the families encountered numerous challenges in the early phase of their displacement. Among other issues, the initial struggles included experiencing hunger, homelessness, stigmatization, and family separation; many families continue to experience these hardships. The family struggles, as exposed through the Nested Model, are illustrated later.

Specific issue

Specific issue of the comprehensive conflict system in the framework of the Nested Model, as illustrated in earlier chapters (Figure 5.1), addresses the immediate issue faced by the people. Similar to those displaced from the Kashmir Valley, the key issue for those who fled from the South Ossetia and Abkhazia regions was being driven out of their ancestral hometowns. This resulted in a feeling of homelessness, followed by the loss of identity, culture, and purpose (Korostelina, 2007), as expressed by one family:

> I was 3-years old when we came from Abkhazia in 1993. My parents had to leave home and become homeless, they brought one bag. My parents travelled together with our extended family and walked for seven days in the mountains. Most of the people came to Samilhone, 200 kilometers from our homes. We brought our shoes and survived because of our shoes. What we have here, we cannot call our home, and we remain homeless.
>
> Abkhazian IDP, now residing in Tbilisi
> (Rajput 2015, personal communication, May 15)

Relational issues

Relational issues of the Nested Model refer to the *fit* of the people into a community, a crucial issue for those having been forcefully evicted who thus must seek shelter into new communities. Although the families eventually came to recognize the members of their host society as "somewhat supportive," their personal and psychological stigma from having lost their bearings adversely impacted their place in the new society. After the prolonged exile, the young children, who were in the early years of their school life when they arrived in new communities, can recall to this day the humiliation of having becoming displaced:

> As a child it was humiliating, it was embarrassing to be identified as an IDP in school, teachers would ask me to stand up and identify myself.
> (Rajput, personal communication, 2015)

This experience of humiliation now comprises this community's "collective trauma" (Volkan, 1997), where the families have marked the day of departure as one of their "most disturbing" memories. Additionally, the families recall the humiliation that came from asking for government assistance as they arrived in their new communities. After two decades, "stigma continues to prevent many members from successfully integrating into the host community and from accessing good-quality [services, such as] education" (Kharashvili, 2012). However, after extensive struggles, the females of this displaced community are beginning to create their own place in the new society. Within the *Tserovani* settlement on the outskirts of Tbilisi, a nongovernmental organization (NGO) known as *For A Better Future*, headed by a displaced community member herself, engages females of the camp in handicraft activities, helping to boost their social and economic status (Rajput, personal communication, 2015). Art pieces produced by these families can be seen in the country's National Museum's gift shop. Initiatives such as these are not only helping the young females of this camp improve their employability but also, more importantly, are serving as a mechanism for them to escape the stigma of displacement, thus making their place into the society more meaningful.

Subsystem issues

Subsystem issues within the Nested Model refer to those "issues that lie underneath the surface of the conflict but those that come to influence the conflict dynamic" (Dugan, 1996), such as the varying value systems or the education systems of the Host and the IDP communities. In cases where there are significant differences in the two communities, those differences percolate to the top and come to negatively impact the displaced families' adjustment. Just as the KP families and the families of the Nagorno-Karabakh region came to be labeled, categorized, and treated as one homogenous group immediately after displacement, the Georgian community also came to be seen as a single homogeneous entity, understood to be needing the same treatment. However, the diversity of origin of these families, their language, and their customs which differed from Georgian mainstream communities impacted their place in new communities. Even with the "somewhat supportive" host community, the children, whose fathers worked away from home and whose mothers struggled to make sense of the Georgian lifestyle in terms of their own hometown traditions, found it difficult to navigate their new society. Unfortunately, these subsystem issues remain outside of the decision-making that factors into IDP policymaking, specifically relating to camp location, camp design, proximity of schools, and the medical facilities provided for the displaced families.

Structural issues

Structural issues within the Nested Model refer to the overarching system of societal rules and regulations that are used to govern a society. To improve the current conditions of the IDPs, Georgia's government in recent times has been able to secure the international support of several agencies, including the UNHCR, USAID, and the Council of Europe (IDMC, 2012; Kharashvili, 2012). However, prior to having secured that support and in the absence of a national approach, "for many years, the IDPs remained marginalized and forgotten and unaware of policies" (Kharashvili, 2012). There continues to be several institutional barriers not only for those still living in exile but also for those who have somehow returned to their hometown communities. Checkpoints, enforced on the administrative borders after 2008, have made it difficult for the families from the Gali district (southern tip of Abkhazia) to cross the administrative boundary to work on their lands. There is also absence of any mechanism for IDPs to recover their housing, land, and property or receive compensation for the losses (IDMC, 2012). Those who remain outside of their hometowns continue to be adversely impacted by the residential segregation, and the compartmentalized communities, as the families are relocated as a group into settlements, which further stifles their integration into mainstream communities.

Relevant policies

The "will to address Georgian IDPs largely came from the support of the NGOs and pressure from civil society" (Kharashvili, 2012). The government's acknowledgment of the IDP *Guiding Principles* in 2000 has also helped to trigger the involvement of the local NGOs within Georgia. Consequently, the country's revised laws now address several IDP issues, such as economic well-being and self-reliance. After 2005, the government also began taking a more systematic approach to the IDP issues, resulting in the 2007 "State Strategy for IDPs" (Rajput, personal communication, May 13, 2015, Ministry of Internally Displaced Persons from the Occupied Territories, Accommodation and Refugees of Georgia). A legal framework now regulates the IDP rights and duties, while the national actors ensure the required training for the officials and promote national awareness of the internal displacement problem. The Action Plan for the improvement of the living conditions of IDPs formulated in 2008 includes measures for housing assistance, livelihood projects, and jobs, (IDMC, 2013). Since 2009, there has also been a move toward a privatization of living spaces and construction of new buildings for IDPs. However, the segregated education, the absence of a mechanism for the restoration or compensation of lost property, and the issues of insecurity continue to pose barriers for durable solutions for the families (IDMC, 2013), as detailed in the following paragraphs.

Housing

The Georgian government has made considerable progress in addressing the housing issues in accordance with its 2007 strategy (IDMC, 2012, 2013), with housing units having been allocated to about 1,500 families. However, transition from the initial collective centers to the new housing has been slow. Consequently, the collective centers, made available for the families during the initial phase of displacement in the city's impoverished areas, with hard to access social services, continue to serve as the main housing for many families (IDMC, 2012, 2013). Many families continue to wait for a housing solution, as many community members remain to be placed to receive such support.

Education

The "self-sustained" and segregated settlements outside of the country's capital of Tbilisi have been a roadblock to schoolchildren's sense of pride as they are perceived to be in the lower socioeconomic ladder of the mainstream. This has hampered the integration of the children and the families. Going forward, children graduating from the segregated school systems also face issues with higher education and job opportunities.

Jobs

Georgia's IDPs have been particularly affected by unemployment (IDMC, 2012). Some unemployed community members are highly educated with professional experience (Tukhashvili, 2010, in IDMC, 2012), while others have lost their professional skills and are no longer able to meet the demands of their new communities (DRC, 5 October 2010, in IDMC, 2012). The officials have put in place some initiatives to ensure availability of jobs for the displaced families. In 2015, a large-scale lemonade factory was opened in the Mtskheta region to employ members of the displaced community of the area (*Agenda.GE*, 2015). However, even with new employment strategies, a large number of displaced families continue to depend on state benefits as their main source of income (IDMC, 2013).

Issue of return

Similar to KPs and the Nagorno-Karabakh families, the issue of return remains the most enduring of the displacement issues for these Georgian families, becoming increasingly challenging for both the policymakers and the families as displacement lengthens. In most cases, the family stance on the issue of return differs significantly from the positions embraced by the officials, as illustrated in the following paragraphs.

Family position on return

The families wishing to return understand that their return is being "blocked" by de facto forces in their hometown communities. These forces fear that a large number of returnees would "disturb the ethnic balance and will compromise security of the region" (IDMC, 2013). Consequently, in recent years, the international response has largely focused on supporting the local integration of IDPs through housing, livelihood projects, and legal programs (IDMC, 2013). Additionally, the families also feel the threat of returning as the "male members of the households that had been black listed by the de facto forces, fear being targeted on return" (Rajput, personal communication, May 13, 2015, IDP of Tserovani camp). Nevertheless, several families have returned to their hometowns under precarious conditions and often to their damaged homes without any assistance in rebuilding. In addition, many returnees have been prevented from accessing their homes and lands, barred by the 50-kilometer fence guarded by the Russian and South Ossetian soldiers. Returnees also remain cautious of investing in order to improve their living conditions, fearing that they may be forced to leave again. In addition, despite the presence of Georgian police in most villages, feelings of insecurity among residents remain high (IDMC, 2012, 2013), and for those who long to return, the issue of security remains their main concern, as expressed in the following sentiments:

> No one wants to go back to unsafe conditions but no one can stop me from dreaming to go back. Here [in Tbilisi] we are not able to live the lives that we had planned on living in our hometown.
> We left Zugdidi [in Abkhazia] on September 27, 1993, they were going to kill us. We don't talk about that day in my family. We were not thinking about the future in another place, only about going back home. My father was black listed, we are not allowed to go back, besides no one wants to go back to unsafe conditions. Here we are not able to live the lives that we had planned and dreamed about but return is tricky.
> (Rajput, personal communication, May 13, 2015, Abkhazian family now residing in Tbilisi)

Official position on return

With the adoption of the State Strategy for IDPs in 2007, the Georgian government's stance on return has shifted from exclusively focusing on return as the only settlement option to supporting the integration of the families at their current locations (IDMC, 2012). The government is also coming to terms with the idea that those displaced from the two regions may not be returning home in the near future. This change in official position has led to an increased spending to improve the current state of affairs of the families. Consequently, the officials are now cautious of recognizing the return as a durable solution, even for those who have somehow managed to return, ruling this as "outside of the organized process" (IDMC, 2013).

Georgian families juxtaposed with Kashmiri Pandit families

A comprehensive examination of the challenges of the Georgian families, as unfolded through the Nested Model, can now be leveraged to build on the current comparative understanding of the displacement of the KP families.

Commonalities

After over 26 years of internal exile, both communities continue to experience a wide spectrum of socioeconomic, political, and psychological issues, such as stigma, family separation, physical ailments, political isolation, and economic hardships. Similar to the KP families, the key issue facing the Georgian families at the time of departure from their hometown communities was the very painful feeling of *homelessness*. The day of departure from their hometowns continues to serve as a "collective trauma" (Volkan, 1997) by both the Kashmiri and the Georgian families. Having lost their bearings adversely impacted their immediate adjustment into their respective host communities. After more than two decades, both communities can recall the humiliating experience of having been forcibly evicted from their own community. As time in displacement prolonged, the loss of ethnic identity, culture, and way of life deepened for both communities. The stigma of displacement continues to prevent the two communities from successfully integrating into their respective host communities.

Having lived in exile for extended periods, both communities continue to depend on government subsidies and assistance in the form of temporary housing, job quotas, and education allocation; such dependence has added to the initial humiliation of having been evicted. The "self-sustained" and segregated settlements outside of Tbilisi, the *Tserovani* complex, for the Georgian families resemble the *Jagti Migrant Township* that houses the KP families outside of Jammu. The segregated settlements have kept both communities isolated from their mainstream communities. This housing arrangement has deprived both communities of the resources required to rebuild their lives and has kept them on the bottom rung of their respective host societies.

Owing to the nature of these displacements, specifically violence induced and protracted, the issue of return remains problematic for both communities as well as for their respective policymakers. Just as the leaders of the Kashmiri families have not yet succeeded in reconciling with the perpetrators of the families or with the state government, similar to the case of the Georgian families, the armed conflicts remain unresolved, and the de facto forces continue to block the return of the families. The perception of danger on return leaves both communities in a predicament on the issue of return.

However, several individualities in the two communities have come to impact the plight of the communities differently, as illustrated in the following section.

Contrasts

One glaring difference in the two communities has been the absence or the presence of external agencies or international actors. Declaring the KP crisis a "sensitive issue," and one that is in the domain of the larger Kashmir problem, served to totally discourage and rule out the participation of external actors, either to investigate the crisis or to provide humanitarian assistance to the families. In contrast, in a recent attempt to improve the conditions of the displaced families, Georgia's government has been able to secure the international support of multiple agencies, such as the UNHCR, USAID, and the Council of Europe. Consequently and under appropriate guidance, the country has adopted a national plan to address the IDP issues, where the international response is focusing on supporting the local integration of the families through housing, livelihood projects, and several legal programs (IDMC, 2013).

1. Official position on return

    While officials of the Indian government continue to maintain the displacement of the KPs as a matter of "temporary crisis," the Georgian government has recently begun to acknowledge that those displaced from the two regions will not be returning home in the near future (IDMC, 2012). This shift has opened space for the authorities to explore context-based durable solutions, such as supporting the integration of the families in their current communities. This has not only lessened the pressure of the officials to pursue return as the only durable option but has also provided the families room and reason to shape their future under this new understanding. This is in contrast to the KP families, who remain in the perpetual dilemma of whether to embrace the new society or wait for their eventual return.

2. Civil society position on return

    The support of the civil society also stands in sharp contrast in the two communities. Whereas there is suspicion among the KP families of the underlying motives and agendas of their own supported groups, where some groups are perceived to have divided the families, the initiative to address the Georgian families has largely come from the support of the NGOs and pressure from Georgia's own civil society. The civil society is also actively engaged in monitoring the implementation of the Georgian government's strategy for the IDPs.

3. Family value systems

    The upbringing and the family value systems may have also contributed to how the two communities came to accept or reject members of their host societies. In the context of the Kashmiri families, the family value systems helped to create the perception that the values of their host communities, such as of Delhi and Jammu, were "diluted" and that their practices of intercaste marriages and nuclear family units represented a "demotion of the KP family values." This had worked to block the wholehearted acceptance of the host communities by the KP families. In contrast, for the Georgian families, disparities of this nature remained beyond the calculus of how families came to regard their counterparts in the host communities, thus facilitating the acceptance by the host and the displaced communities of each other.

Conclusion

Overall, there continue to be structural and institutional barriers that impede the growth of the displaced families in both the Kashmiri and the Georgian communities. The return of the communities to their respective hometowns, if it is to materialize, would leave the families with inadequate mechanisms to recover their land and property. The ancestral properties for both communities have been illegally occupied by militants in the Valley or the de facto forces in Abkhazia and the South Ossetia regions. Those who continue to remain outside of their hometowns may continue to yearn for the life and the livelihood they were forced to leave behind while continuing to experience the segregation and humiliation that comes from the labeling of their communities and dependencies that come from state-provided subsidies, while they remain in extended exile within the borders of their home countries.

The global journey into conflict-induced displacement continues in the next chapter, with a methodical examination of the challenges of the ethnic Serbs who fled from Kosovo in 1999 and remain displaced within Serbia after almost two decades, falling within the rubric of the Serbian IDPs. The exploration of the displaced ethnic Serbs is a further attempt to position the displacement of the KP community in comparative perspective.

References

Agenda. GE. (2017). Multi-million GEL lemonade factory opens in Georgia. Retrieved from http://agenda.ge/news/38893/eng

Dugan, M. (1996, Summer). A nested theory of conflict. *Women in Leadership, 1*(1), 9–19.

IDMC. (2012, March 21). *Partial progress towards durable solutions for IDPs.* (Summary of IDMC's internal displacement profile on Georgia). Geneva, Switzerland: Norwegian Refugee Council.

IDMC. (2013, December 31). *Georgia: Internal displacement in brief.* Geneva, Switzerland: Norwegian Refugee Council.
Kharashvili, J. (2012, November). *20 years of internal displacement in Georgia: The international and the personal.* FMR's 25th Anniversary Collection.
Korostelina, K. (2007). *Social identity and conflict: Structures, dynamics, and implications.* New York, NY: Palgrave Macmillan.
Volkan, V. (1997). *Blood lines: From ethnic pride to ethnic terrorism* (2nd ed.). Boulder, CO: Westview Press.

12 Serbia
Displaced from Kosovo (ethnic Serbs)

Introduction

The mission to advance an understanding of the Kashmiri Pandit (KP) displacement through a comparative lens continues in this chapter, with a systematic exploration of the challenges of the displaced ethnic Serb community. The ethnic Serbs who fled from Kosovo in 1999 remain displaced within Serbia after almost two decades. At the outset, the displacement of ethnic Serbs ousted from the neighboring Republic of Kosovo may seem unfitting to gain a comparative understanding of the internally displaced KPs. Although Kosovo was admitted as an Autonomous Province of the post-World War II Yugoslav Federation, Serbia remains the only independent country now as Kosovo declared independence from Serbia in 2008. However, Serbia has not recognized Kosovo's independence and treats the border between Kosovo and Serbia as an "administrative line" under shared control. Consequently, the ethnic Serbs who fled from Kosovo and have been living in exile within Serbia fall under the rubric of those "internally displaced" within Serbia as opposed to refugees who have fled from the neighboring country. As of the end of 2014, the Serbian Commissariat for Refugees (SCR) reported that around 204,000 of those who fled from Kosovo in 1999 remain within Serbia (IDMC, 2015). Under this understanding, the displacement of this ethnic minority displays essential similarities to the ouster of the KP community, relating to the cause of displacement, magnitude of the exodus, time in internal exile, and level of human suffering inflicted on those displaced.

The Republic of Serbia is situated in Europe's Balkan Peninsula at crossroads between Central and Southeast Europe. With Belgrade as its capital, Serbia has a population of seven million, with a tiny minority of ethnic Serbs still living in Kosovo. In recent years, Serbia has become a transit country for those fleeing many parts of the world embattling the violence-induced displacements from their own countries. The final destinations for many of those transiting into Serbia are other parts of Europe. Historically, as part of the Socialist Federal Republic of Yugoslavia (SFRY), Serbia had been managing migration flows for some time. Consequently, from 1992 to 2000, the Serbian government "admitted one million people from Croatia, Bosnia and Herzegovina, as well as IDPs from Kosovo" (personal communication, May 6, 2016, official, Serbian Commissariat for Refugees), thus becoming host to the largest number of refugees and IDPs in Europe.

Circumstances of displacement

In 1989, Slobodan Milošević, at that time the President of SFRY, campaigned for a reduction of powers for the various autonomous provinces under his domain, including the autonomous province of Kosovo. This activated a wave of nationalism across the Federation, resulting in the breakup of the Federation and the consequent independence of the separate republics. In 1999, consequent to the actions of the North Atlantic Treaty Organization (NATO) that forced the Yugoslav troops from Kosovo (IDMC, 2014), about 245,000 ethnic Serbs, fearing revenge from the majority Albanian population, fled their homes in Kosovo. The arrival of those displaced combined with the number of refugees already in Serbia made the country host to one of the largest populations of displaced people in Europe.

Challenges encountered

After almost two decades, the main needs of the displaced Serbian families revolve around housing, recovery of lost property, employment, and the right to return to their homes in Kosovo. The nature of challenges encountered by the families immediately after their displacement and those that continue to overwhelm the families, as framed around the Nested Model (Dugan, 1996; an exhaustive understanding of the Nested Model is detailed in Chapter 5), are discussed later.

Specific issue

Specific issue of the comprehensive conflict system in the framework of the Nested Model addresses the immediate issues faced by the people (Figure 5.1). As the crisis deepened in Kosovo in 1999, the ethnic Serb minority, fearing revenge from the majority community, fled to neighboring Serbia. This unplanned separation from their ancestral hometowns was the key concern of the families as they fled. The Serbian authorities maintain that the permanent residence of those who fled remained in Kosovo and accordingly issued "temporary residence identity cards" to those arriving at the various *Collective Centers* in Serbia. However, having been forced to abandon their homes and deprived of the recognition of being "Kosovo's people," people immediately experienced a deep loss of identity (Korostelina, 2007) and a loss of purpose. Since the arrival of the first families in 1999, the Centers have seen expansion of the families with the arrival of relatives and many newborns. Like the identity of those who fled, the identity of these newborns remains in limbo.

Relational issues

Relational issues of Dugan's comprehensive conflict system refer to the *fit* of the people into a community. Serbian officials mentioned that those who arrived at the Centers "easily settled in the Serbian communities, without any issue of adjustment, as they were already a part of the Serbian culture, from former Yugoslav territories …they are [our] country-men" (personal communication, May 6, 2016, Serbian Commissariat for Refugees). The officials view assimilation as "more of a thorny issue for refugees arriving from other countries" (personal communication, May 6, 2016, Serbian Commissariat for Refugees). However, in search of jobs and better access to services, during their prolonged displacement, many families began to move out of the Centers to larger cities within Serbia. Consequently, many came to settle in Belgrade; this move put heavy pressure on the families, forcing them to cope with the metropolis and to navigate the bureaucracies of the big city, which continues to hamper their adjustment into these communities.

Subsystem issues

Subsystem issues within Dugan's Nested Model refer to those issues that lie underneath the surface of the conflict but percolate and come to influence the conflict dynamic. In the context of the displaced communities, such issues come to impact the IDP/Host dynamic, governing their day-to-day interactions (Rajput, 2012). Being part of the former SFRY, both Kosovo and Serbia followed the same school systems. This facilitated the transition of the children of the displaced families into their local schools immediately upon arrival into host communities. Consequently, there are no dedicated schools on the camp premises or any segregated schools for the children of the displaced families within the mainstream communities. In addition, as Serbia has been a transitioning country for the many refugees on their onward journey to Europe, the locals have been accustomed to seeing the families from diverse backgrounds moving in and out of their localities. Therefore, the arrival of the ethnic Serbs from Kosovo did not raise any flags or generate hostile reaction from members of the local communities, which in most other cases of internal displacements quickly leads to the divide and the advent of two parallel communities.

Structural issues

Although the schoolchildren may not have experienced major changes in the school systems, the adjustment of the families was hampered by the many institutional challenges that continue to be imposed on them, such as those required to prove their eligibility for the benefits. The lack of personal documentation has been a recurring bottleneck for the families, limiting their access to employment, medical assistance, and other social benefits. After more than two decades, the families continue to be challenged to meet their everyday needs and obligations.

Relevant policies

The Serbian Commissariat for Refugees that oversees the issues of the country's refugees and those who remain internally displaced within the country as well as the issues of the minority communities within Serbia is recognized as one of the most important agencies in Europe. The government has instituted a *National Strategy for Resolving the Problems of Refugees and Internally Displaced Persons* (R&IDPs; SCR, 2002) with specific goals that address employment, education, health insurance, and housing. The following paragraphs detail the key policies that are currently extended to the displaced families.

Housing

As an indicator of a step toward durable solutions, the government's strategy, supported by the European Union (EU), is to reduce the number of the current beneficiaries of the many Centers and secure alternative housing solutions for the families. Currently, there are 900 people living in the Centers, substantially reduced from 63,000; "the plan is to use the existing Centers to accommodate the transitioning refugees and asylum seekers from parts of Africa, Afghanistan, and elsewhere" (personal communication, May 6, 2016, Serbian Commissariat for Refugees). Authorities are trying to establish special sections of prefabricated housing or village houses to house the displaced families currently residing at the Centers. Despite these efforts, for over two decades, close to half of the registered IDPs continue to live in precarious conditions, with a large number of families having remained in the areas to which they were initially displaced, living with limited access to basic services.

Education

The officials maintain that as the children of the displaced families share the culture and language of the larger Serbian community, this has "ruled out any need for special education policies or special rights, for those children" (personal communication, May 6, 2016, official, Serbian Commissariat for Refugees). However, Serbia's Ministry of Education has initiated a number of reforms, which includes addressing the advancement of the IDPs and other vulnerable social groups. As a result, many IDPs have participated in the "Second Chance" program, which allows people to complete their elementary education (IDMC, 2013).

Jobs

There are no specific job policies or dedicated training for the IDPs living at the Collective Centers. Consequently, the IDPs who live in social housing are mostly unemployed or retired, or receive unemployment benefits. The officials suggest that many of the displaced living in the Centers have become so dependent that they are "pleading the officials to keep the Centers open" (personal communication, May 6, 2016, official, Serbian Commissariat for Refugees). Additionally, the officials fear that as the people have been living at the Centers for over 20 years, they may no longer be competent enough to survive outside the Centers (personal communication, May 6, 2016, official, Serbian Commissariat for Refugees). Consequently, the IDPs, many of whom have been unemployed since their displacement (IDMC, 2013), continue to suffer from a higher rate of unemployment compared to the general population. IDMC reports that a 2013 UNHCR survey revealed that 32% of IDPs were unemployed compared to the 19% national unemployment rate. While the government continues to extend efforts to improve the families' access to livelihood projects, the efforts have yet to achieve the intended impact of improving the families' level of self-reliance.

Issue of return

Similar to the sentiments of the displaced communities discussed earlier, such as those displaced from Azerbaijan's Nagorno-Karabakh regions, those displaced from Abkhazia and the South Ossetia regions of Georgia, as well as the KPs displaced from the Kashmir Valley, the issue of return remains the most troubling for these Serbian families. The hope of return seems even dimmer as the government initiative to close the Centers and move the people into prefabricated housing becomes more real. The following paragraphs highlight the various positions embraced by different actors on the issue of return.

Family position

Voluntary returns by the Serbian families to their place of origin in Kosovo have been scant. After the lengthy time away from home, IDMC (2013) reported that only around 3% of those displaced had returned to their home communities, with the sustainability of those returnees remaining doubtful. The families remain skeptical of their return, fearing that as a minority community, they may continue to be treated as such, barred from access to their property and deprived of the needed resources.

Official position

Political participation, in the case of displaced and otherwise divided communities, represents a solid marker toward a commitment to durable solutions. Kosovo held its municipal elections in 2013, which translated into new hope for those who remain displaced across the border. The elections gestured increased political and ethnic representation of the ethnic Serbs and a promise of improved social conditions if the families were to return to their ancestral towns (Turner & Walicki, 2014). Kosovo's elections have offered the families a chance to be heard; to regain their lost rights; and, most importantly, to be counted (IDMC, 2013). Consequently, the families are in a better position to weigh their options, given the cooperation between the governments of Belgrade (Serbia) and Pristina (Kosovo). In addition to the return to home communities, the local government within Serbia has prepared action plans to assist the families with local integration at their current residence. However, the Serbian officials make it clear that as "there is no specific government return policy, family return is based on the family's own wishes and is honored, based on a culture of compassion and care" (personal communication, May 6, 2016, official, Serbian Commissariat for Refugees). However, the officials remain doubtful as to whether the families will be comfortable returning to their hometowns as a minority community and are beginning to accept this as a "reality of the prolonged" displacement of this ethnic minority community.

Serbian families juxtaposed with Kashmiri Pandit families

The account of the challenges of the Serbian families, as exposed through the Nested Model, shows a number of commonalities as well as significant contrasts between the KP community and the ethnic Serbs, as elaborated upon in the following.

Commonalities

Despite the perceptible difference between the Kashmiri and the Serbian communities – one having been displaced from an autonomous province that later became an independent sovereign Republic of Kosovo and the KP community that remains displaced within its own sovereign nation – the two communities were made to endure the same hardships that resulted from being evicted from one's own society. The minority status of the two ethnic communities, their displacement stemming from a wave of nationalism, and the extended time in exile also puts the two communities on a comparable scale of human suffering. After almost two decades of exile, members of both communities remain preoccupied with aspects of rebuilding their lives, such as securing suitable housing for their growing family; securing sustainable employment; coping with the Host/IDP differences; and, most painfully, living with an uncertain future.

Similar to the KP families, the specific issue facing the ethnic Serbs, as they were fleeing Kosovo and separating from their ancestral homes, was the feeling of having become *homeless*. As the families were leaving, what may have taken them a lifetime to build was being handed over to their perpetrators. The ethnic Serbs were being deprived of their recognition as "Kosovo's people" just as the Kashmiri families were being deprived of being the "Pandits of the Kashmir Valley," eroding the character and the uniqueness of their very being (Avruch, 2011).

Contrary to the common perception that being a part of the larger Serbian and Indian cultures may have facilitated the adjustment of these two internally displaced communities into their respective host societies, during their extended exile and in pursuit of economic survival, families of both communities moved out of the *Centers* and the *Migrant Townships* to metropolitan cities. The arrival into the larger cities of Belgrade and Delhi put tremendous pressure on members of these communities, stifling their adjustment into the urban locales. Since the arrival of the first families, the *Centers* and the *Migrant Townships* have witnessed the expansion of the communities, with the addition of many newborns. Together with the families, the identity of these children also remains "temporary" as the authorities in both cases steadfastly hold to the position that the permanent residence of these displaced families remains in their hometowns of Kosovo and Kashmir.

Similar to the sentiments of the KP families, the issue of return remains the most troubling for those displaced from Kosovo. The hope of return becomes dimmer as the government initiative to close the Centers and move the people elsewhere becomes the new priority. This mirrors the sentiments of the KP families, who sensed the building of the *Jagti Township* on the outskirts of Jammu as an indication of the "government's failure to reconcile with the families' perpetrators and the government's inability to secure safe return to the Valley" (Rajput, 2012). Additionally, voluntary returns by the families to place of origin have been scant in both communities, with the families remaining wary of the security that can guarantee their safe return and keep them protected in their hometowns. The presence of security forces in the Valley as well as in the surrounding areas within Kosovo instill the perception of fear in those wanting to return to their home communities. The families also fear retaliation by their perpetrators and fear being labeled as unfaithful by those who stayed in their hometowns. The families also suspect that as minority communities, they may continue to be treated as such, remaining on the bottom ladder of their respective societies.

Contrasts

A unique feature of a national-level crisis taking place within any of the European countries is that it comes to garner regional and often international attention and interest. This is also true of human displacement that may stem from any of the triggers, such as conflict-induced, natural disasters or the government's own development plans. This very feature translates into external intervention, financial aid, and coordinated material and technical support, including assistance from the international community. This aspect stands in stark contrast to how the displacement of the KPs had come to be recognized and subsequently handled. The following paragraphs address key differences in the two cases, given the presence/absence of external or international actors in response to internal displacement.

The Serbian agency (SCR) that oversees the IDPs within the country also enjoys an international reputation as one of the most significant agencies in Europe. The specific goals that address the needs of the displaced ethnic Serbs are aligned with the priorities of the EU, which calls for the closing of the Collective Centers in pursuit of more durable solutions for these families. This is in sharp contrast to the plight of the KPs, which has remained a national issue, governed by the national discourse and remaining on the margins of the countrywide priorities. In the national discourse, the issue of the KPs either remains in the domain of "sensitive issues" or as a "settled issue," and thus outside of the sphere, concerns, or interests of external actors. The Serbian effort largely supported and guided by international actors has been able to put in place a national strategy with a focus on housing, jobs, and education for the country's displaced communities, whereas the country's national leaders have ruled out any form of international support for the displacement of the KP community.

It is also important to understand that Serbia, being a transitioning country for those fleeing violence from other parts of the world, has familiarized the locals of the random movement of people into their local communities. Consequently, the arrival of the ethnic Serbs from Kosovo did not generate excessive negative reaction and hostilities among the locals of Serbia, whereas the arrival of the KPs into host communities of Delhi, Jammu, and elsewhere created an immediate uproar that resulted in the division of host communities. The influence of the locals also became a strong driver in the shaping of the KP policies. As revealed earlier, SCR has been working with international partners on better housing solutions, whereas any housing solutions for the Kashmiris remain bogged down by the issue of return, which remains problematic, given the conflicting positions taken up by the KP family advocates themselves.

However, unlike those extended for the Kashmiri community, there are no specific job policies or special quotas for the ethnic Serbs from Kosovo. Consequently, those who have been living at the Centers are mostly unemployed and now fear being kicked out without job-related competencies to support them. In contrast, the job policies that came through such initiatives as the "temporary use of the shops" and the job quotas through the *Prime Minister's Special Allocation for the Migrants* has equipped the KP community with lifelong and employable skills, which has ensured the community's survivability outside the *Migrant Townships*.

Through the years, Serbian officials have maintained the position that given the fact that the children of the displaced families were already familiar with the national system of education, there was no need to grant the families distinct rights and education quotas or set up segregated schools. This is in contrast to the education programs offered for the children of the Kashmiri families. Initially, the dedicated schools for the Kashmiri children were set up on the camp premises, in practice instituting a segregated schooling for the children, which later came to stigmatize them as they transitioned into the larger mainstream schools. The entry to the host community school systems came through a "special allocation for the children of the migrant families." As the Kashmiri children transitioned into these new schools, they were tasked with the challenge of adjusting to the rules of their new and unfamiliar school systems, whereas the children of the Serbian families were spared this hardship.

Conclusion

After more than two decades, the many challenges faced by the displaced KP and the ethnic Serbian communities, specifically attributed to their displacement, continue to impact many aspects of their daily lives as both communities continue to live with diminishing prospects for return. Even if the promise of safe return and sustainable security can be guaranteed, the mistrust for their respective authorities, who are perceived to have allowed their displacement to take place, may pose the strongest barrier to the families' eventual return to their hometowns.

The quest to deepen an understanding of internal displacement continues in the next chapter, with an examination of the challenges of those displaced within Sudan, where the ongoing conflict in Darfur had displaced close to three million people.

References

Avruch, K. (2011). *Context and pretext in conflict resolution: Culture, identity, power, and practice.* Boulder, CO: Paradigm Publishers.

Dugan, M. (1996, Summer). A nested theory of conflict. *Women in Leadership, 1*(1), 9–19.

IDMC. (2013, December 31). *Georgia: Internal displacement in brief.* Geneva, Switzerland: Norwegian Refugee Council.

IDMC. (2014). *Azerbaijan: After more than 20 years, IDPs still urgently need policies to support full integration.* Geneva, Switzerland: Norwegian Refugee Council.

IDMC. (2015). *Serbia IDP figures analysis*. Geneva, Switzerland: Norwegian Refugee Council.

Korostelina, K. (2007). *Social identity and conflict: Structures, dynamics, and implications*. New York, NY: Palgrave Macmillan.

Rajput, S. (2012). *The displacement of the Kashmiri Pandits: Dynamics of policies and perspectives of policymakers, host communities and the Internally Displaced Persons* (Doctoral dissertation). Fairfax, VA: George Mason University. ISBN: 9781267843333.

SCR. (2002, May 30). *National strategy for resolving the problems of refugees and internally displaced persons*. Government of the Republic of Serbia, Belgrade.

Turner, W., & Waliciki, N. (2014). *IDP rights in Kosovo and Serbia should be a key consideration for EU membership talks*. Geneva, Switzerland: Norwegian Refugee Council.

13 Sudan
Displaced from Darfur

Introduction

An effort to unveil the plight of those who remain as internally displaced persons (IDPs) having fled from the Darfur region of Sudan is intended to enhance a comparative understanding of the ethnic minority community displaced from the Kashmir Valley, the Kashmiri Pandits (KPs). The ongoing conflict in Darfur that began in 2003 displaced several million people. Subsequently, Darfur has been plagued by overlapping layers of violence, rebel group fighting against government forces, irregular militias fighting against rebel groups, and nomad groups fighting other nomad groups over limited resources (Barakat, 2014).

Darfur is located within the Republic of Sudan in North Africa and occupies the Western region of the country. Sudan is bordered by Egypt, Ethiopia, South Sudan, Chad, and Libya. The country is recognized through a range of ethnic, religious, cultural, climatic, and livelihood diversities, which has made the country a small-scale version of Africa itself. However, Munzoul (2004) explains that in the international context, the Sudanese are often placed into much-simplified twofold categories on the basis of the country's geography (North versus South), religion (Muslims versus Christians), and ethnicity (Arabs versus Africans). Having remained an autonomous precolonial state for several centuries, the entire Darfur region, covering an area of about 500,000 square kilometers, was incorporated into the country of Sudan in 1916. The four centralized areas of the Darfur region continue to be East Darfur, West Darfur, North Darfur, and Central Darfur.

Circumstances of displacement

In 2003, Darfur became the scene of violence when non-Arab groups within Darfur took up arms against the military campaign of the

Sudanese government, responding to the government's oppression of non-Arab Sudanese in favor of the Sudanese Arabs, triggering a civil war in Darfur. Some studies explain the Darfur crisis as rooted in the communal disputes over natural resources such as water (Salih, 2008). Faced with consistent drought lasting for several decades, the local nomadic tribes left their homes in search of water and ventured into new territories. The hostility inflicted by members of new territories to the outsiders led to clashes between the farmers and the nomadic tribes. Combined with the recurring violations of the basic human needs of Darfur's people, the abundance of weapons in the region and the lack of economic development resulted in a full-scale civil war in 2003 between the Sudanese government and several Darfur rebel groups (Rothbart, Brosche, & Yousif, 2012). Since 2003, the violence in Darfur has led to the deaths of 300,000 people (Campo, 2016) and has displaced more than three million people (World Without Genocide, 2013). A large number of people are being housed in several camps, with *Abu Shouk* serving as the largest of the camps, located 4 kilometers north of El Fashir, capital of North Darfur State. IDP camps in Darfur remain under the authority of the Sudanese government and the United Nations African Mission in Darfur (UNAMID).

Challenges encountered

Unfortunately, when conflicts linger for decades, and other priorities take over, the issues of those who remain displaced get lost or bogged down, while simultaneously, new challenges add on to complicate the plight of those who remain displaced. The original conflict that began with the rise of the non-Arab tribes in Darfur against the government now has added components of territorial issues, cattle-raiding, and issues of ethnic and tribal identities. Consequently, Darfur families face multiple and evolving challenges personally, socially, economically, legally, and politically. Those living near bordering countries, such as near Chad and South Sudan, are also exposed to the hostilities of their neighboring countries, countries that are themselves prone to underdevelopment and poverty. Relying on the analytical approach that has unfolded the challenges of those displaced from the Kashmir Valley (Rajput, 2012), Azerbaijan, and Georgia as well as the ethnic Serbs displaced from Kosovo (discussed in the preceding chapters of this book), issues of Darfur's IDPs can also be better understood through Dugan's Nested Model (Dugan, 1996). In the context of Dugan's Model, the issues of a *specific, relational, subsystem*, and *structural* nature (Figure 5.1) that confront Darfur families are systematically scrutinized later.

Specific issue

In Dugan's framework, the *specific* issue refers to the immediate issue facing the conflictants (ref. Chapter 5). For the Sudanese people, the family unit represents one of the most honored social institutions. In addition to its basic role of human reproduction and the upbringing of children, the family is where the "vulnerable segments of the population, such as the elderly, [belong] and [feel] at home" (Munzoul, 2004). As the families fled the violence-infected areas of Darfur in search of safety elsewhere, they suffered the major breakdown of this time-honored tradition of family unity. Many families secured shelter outside of Darfur with family members scattered in varied parts of Sudan, thus destroying the family structures. For Darfur families, having been uprooted from homes and unable to harvest crop and tend to animals meant a complete termination of the only lives they knew. The *specific* issue for the families as they fled was the loss of home and kinship. Loss of home also meant the loss of functions that were traditionally performed by family members, such as providing for and taking care of the elderly and the young. The alternative of having become the recipient of welfare and handouts was a shameful trade-off for these families. The loss of home led to complete erosion of people's sense of social worth (Korostelina, 2008).

Relational issues

Relational issues in Dugan's Model refer to the *fit* of the people into a community. Once displaced, the families often find themselves socially, politically, and culturally in an unwelcoming and hostile environment. The arrival of Darfur families in host communities was directly affected by the sociocultural perspectives of their host societies, which embraced negative conceptions of the displaced, associating them with security and health issues, and positioning them as disrupters to the societal order of their communities. Their host "community's negativity" eventually pushed the families out to the suburbs of Khartoum (Munzoul, 2004). In new surroundings, within the confines of the government-controlled camps, the families were confronted with the daunting task of recreating and managing their new life patterns. Although the IDPs in Darfur generally came from the same region, a shared common religion, and a shared culture (Avruch, 2011), the historical rivalries between the groups stemming from competition over resources (Rothbart, n.d.) stifled their initial adjustment into the camps. Thus, the *Fur* and *Zagawa* tribes identified themselves differently, despite their shared religion and culture

(Korostelina, 2008), which led to hostile relations and conflicts inside the camps. Consequently, adjustments even within the camps became a challenge for these displaced families.

Subsystem issues

In the context of the Nested Model, the *subsystem* issues refer to those issues that remain underneath the surface of the conflict but have the potential to rise to the top in the event of an uptick in the conflict. Regardless of the basic overarching religion, most Darfur families followed their tribal and ethnic identities, which clashed with their religious and group identities (Rothbart, n.d.). In addition to the adjustment challenges within the camps, the families were challenged in adapting to the values and cultures of members of the mainstream communities. The core values of the Darfur families were rooted in traditional agriculture and animal husbandry; however, these nomadic values were not on par with what the host communities valued, thus exposing the families to a "cultural milieu" alien to them (Munzoul, 2004). Consequently, the families' personal attributes, such as their coping skills, parenting skills, upbringing, and moral ethos positioned them as alien, strange, and out of place in their mainstream communities.

Structural issues

Structural issues of a comprehensive conflict system refer to the overarching systems for rules and regulations that govern a society but that often come to block the actors' access to the needed resources. As exposed in earlier chapters, the issue of displacement is never an isolated phenomenon, but rather falls in the domain of complex conflict issues. The displacement of the large number of people stemming from the instability in the Darfur region, suggests the malfunctioning of the overall state institutions. As the non-Arab Darfur population historically remained marginalized within the larger Republic of Sudan, the massive displacement of this community resulted in their continued victimization by the state-run policies, keeping them on the fringes of the larger national society.

Despite the effort of the international community to keep the IDP camps secure through UNAMID, the security of residents has not been always guaranteed; women have encountered security issues, suggesting that their security remains a low priority. In addition, aid agencies have faced obstacles in carrying out their own missions (Newvision, 2014). In 2014, several aid agencies, including the International Committee of the Red Cross, were suspended from the Republic of Sudan. The essential services of sanitation and water also remained overly overstretched. As the root of Darfur conflict remains embedded in the policies that inflict political and economic marginalization, the displaced families continue to remain victims of the "institutional violence" (Galtung, 1969) entrenched in the structures of the society.

Relevant policies

Given the nature of prolonged displacement, where the initial issues confronting the displaced families often come to be embedded and overlapped into other issues, the policymaking for the Darfur families remains complicated. In large part, the care and responsibility for the displaced families has been delegated to various nongovernmental agencies (NGOs), thus relieving the officials of the tricky task of IDP policy formulation. Nevertheless, the lawmakers face a dilemma as to how best to address the issue of return and other essential policies that impact the families' basic survival. The families understand the "current system of relief to be of a short-term nature, being pursued at the expense of long-term solutions" (Osman & Sahl, 2000, pp. 4–5). The key policies relevant to the displaced Darfur families are explained later.

Housing

The housing options for Darfur families essentially relate to residing in one of the many camps in and around Khartoum. The official perception of the displaced families as a "security threat" (Munzoul, 2004, p. 7) has required that camps be located away from the mainstream communities. The embracing of this narrative has not only challenged the families but also, more importantly, restricted the access of the humanitarian agencies, stifling the delivery of essential services for the families residing in the camps. These security-fearing policies have closed the space to secure context-based solutions for the families. However, in a recent attempt to integrate the families, the authorities have begun to explore formal integration of the families through allocation of residential property (Munzoul, 2004, p. 7).

Education

The families' prolonged captivity in displacement camps has taken away any opportunities of education for the youth. However, a glimmer of hope has emerged from the works of the local NGOs, such as the *Mujaddidon Organization*, which has recently set up an open street school for the "homeless or the street children" (Huaxia, 2017), including the victims of the Darfur displacement. The program focuses on rehabilitating and changing the negative behavior of these children while educating them with school curriculum. The children are being attracted to the school in exchange for meals that they desperately need, as some children live in the abandoned buildings and tunnels within and around Khartoum.

Jobs

The remote location of the IDP camps has kept Darfur families cut off from the mainstream and deprived of engaging in activities that would boost their economic well-being. This has made any job opportunities virtually nonexistent for the families. Consequently, the heavily populated camps, such as *Al Salam* are relegated to the category of absolute poor (Osman & Sahl, 2000) as these families are understood to be unable to survive on their own if they were to leave their camps. However, the families want to contribute to the development of their new communities as well as their hometown region as opposed to living on state subsidies indefinitely.

Issue of return

The issue of return remains delicate for most communities in protracted displacements, complicated by multiple variables, such as the time in displacement, instability of their hometowns, collapse of livelihood, lack of trust, and the issues of generation gap. The idea of return is also complicated when the conflict remains unresolved, as in the case of Darfur. Consequently, for Darfur families, the willingness to return requires an assessment of myriad issues, such as the restoration of land rights; prospects for livelihood; securing of the community's trust; and, most importantly, family's own sense of security. In addition, the different *positions* (Harré & van Langenhoven, 1999) that are being embraced by the families as well as by the policymakers also weigh heavily on the prospects of return, as typified in the following sections.

Family position

The ongoing conflicts in Darfur have destabilized large parts of the hometown communities, including the traditional means of livelihood. The absence of alternative skills and education while in exile puts a heavy burden on the families, who long to secure sustainable means of survival. These factors contribute to the dilemma that the families face on the issue of return; additionally, the families find themselves puzzled, not knowing who to trust, who to turn to as the members of civil society are also isolated and underrepresented. The families express the humility that comes from the government subsidies and their anxiety of return as follows:

> Definitely we want to return, it is illogical to live in the camp forever. We want to return to our villages and our works of cultivation and grazing. We do not want to depend on what is provided by the organizations of food it does not meet our needs.
> (IDP of Abu Shouk camp in Xuequan, 2015)

Darfur families also face a unique challenge on their return to their hometown communities. The division of Darfur into the new five administrative units, with the addition of "central Darfur" and "eastern Darfur" to the previous three regions poses a challenge of a different kind for the returnees. This setup increases the possibility for new conflicts as the rival groups may find themselves in the same administrative unit, raising the possibility of one group dominating the other. Under this scenario, the people who were part of the majority state may now fall under the minority, and vice versa: for instance, the *Rizeygat Arabs*, in their new state, have become a majority, thus opening the doors for new conflicts (Rothbart, n.d.). The advent of the new administrative units also complicates the restoration of land rights as land and property may now fall under another party's control.

Finally, as communities have their own perception of threats, willingness to return is also entrenched in the perceived level of security; the returnees want security on their own terms. The presence of armed rebel groups conflicts with the idea of returning to a peaceful place.

Official position

At one time, issues of the Darfur people occupied first place in international peace and security discussions; however, in current times, the status of Darfur's displaced people has lost the same visibility. Nevertheless, in 2014, the Sudanese government initiated the process of a national dialogue, encompassing intensive discussions on issues of identity, human rights, and peace. Implicit in those discussions was the idea of return of those who remain displaced from Darfur. The conference brought together 700 participants in 2015 at the government's headquarters in Khartoum (author's participation at conference, October 2, 2015); however, many key armed groups and members of opposition that must be involved for any resolution to be sustainable refrained from participating (Campo, 2016). Pending successful peace agreements with multiparty actors, including the perpetrators responsible for ousting the families, the officials are now working with the idea of integrating the families within their host communities.

Civil society position

The proposed national integration policy has led to some confusion within the IDP families (Munzoul, 2004). Although the government's message signals integration through residential plots, several NGOs instead see voluntary repatriation as a priority. Consequently, plans for alternative to integration are taking place at the grassroots level. In several Darfur areas, the Arab nomads have organized conflict resolution networks to address local disputes (Barakat, 2014). Pending political agreement, other grassroot efforts involve developing the area of *Al Malam* by rebuilding schools, hospitals, and markets with the hope that more people will return and begin to improve their livelihoods (Campo, 2016). *Abu Shouk*, the largest of the IDP camps in Darfur, is also being restructured into residential cities, to become a part of El Fasher city, with essential services, which will allow the displaced families to become a part of the community (personal communication, October 2, 2015, Darfur IDPs, in Khartoum). However, half-hearted attempts on the part of either the government or the NGOs could result in the failure of all three options of integration, repatriation, and relocation (Brookings-Bern, 2008), leaving the IDPs in limbo.

Darfur families juxtaposed with Kashmiri Pandit families

The interpretation of the challenges of Darfur families, as framed around the Nested Model, can now be leveraged to position the understanding of the displaced KP families in comparative perspective. Given the nature of the two displacements, both communities continue to face similar challenges of a personal, social, economic, legal, and political nature. The following paragraphs illustrate the commonalities as well as the individualities of the two displaced communities.

Commonalities

To a large extent, the resemblance of the two communities can be understood as comprising a range of issues, which can be explained under the overarching category of "broader issues" and the specific issue dedicated to the "issue of return":

1 Broader issues

Having remained an autonomous precolonial state for several centuries, the region of Darfur was later incorporated into the country of Sudan, bearing resemblance to the autonomous nature of Kashmir that had remained a "princely state" under the British Empire and was later incorporated in the sovereign nation of India in 1947. Within their respective Republics, both communities remained the minority communities and often, on that account, faced injustices on many fronts, such as unfair distribution of resources and unequal political participation, eventually becoming the scene of violence that led to the displacement of a large number of their community members from their homes. In many cases, the families in both communities of the Valley and Darfur were moving out of their ancestral homes for the first time.

Given the protracted nature of both displacements, several family issues have been lost or bogged down while new challenges have added to the continued predicament of the families. The original conflict that began with the rise of the non-Arab tribes in Darfur against the government and the rise of militancy against the Delhi government in the Valley now have additional dimensions, complicating the resolution of respective conflicts. The issues of territorial boundaries, cattle-raiding, and preserving the essential *Kashmiriyat* character of the Valley, which came from the amalgamation of the diverse ethnicities that once inhabited the Valley, have now added to the complexity of the original issue, responsible for the ouster of these two distinct communities.

For both communities, the family unit represents one of the most valued social institutions. For Darfur families, being uprooted from homes and unable to harvest crop and tend to animals meant a complete disorientation of the only lives they knew. Kashmiri families experienced a similar void, having been deprived of the responsibilities to attend to their walnut orchards and care for the farm animals (Rajput, 2012). The *specific* issue for both displaced communities as they fled their hometowns was the loss of home that entailed being removed from the symbols of worship and community practices, and having been robbed of a unique identity (Avruch, 2011). Although the overarching culture and religion of both communities fell under the shared practices of their larger host communities, having arrived with the IDP and the migrant labels, the families were challenged into demonstrating their loyalty and *fit* into the new societies. However, Darfur families came to be associated with security and health issues, while the Kashmiri families were associated with being dishonest and lazy in having exploited the government programs. The protests from the locals obligated the leaders of both groups to seek housing solutions for the families on the outskirts of the mainstream host communities of Khartoum and New Delhi. Displacement of both communities has exposed the breakdown of the governing systems in their respective societies. Just as a general practice, the non-Arab population had remained sidelined within the larger Republic of Sudan; similarly, the ethnic KP minority had remained unprotected within the majority Muslim society, which allowed the displacement and the denigration of this community.

2   Issue of return

Similar to the sentiments of the Kashmiri families, the issue of return remains a painful yet delicate issue for the Darfur families as the conflict remains unresolved in both Darfur and the Valley. Both communities weigh prospects of return in terms of the time outside of home communities, instability of their hometowns, restoration of property, collapse of structures, fragmentation of the community, fear of retaliation on return, and most importantly, the lack of trust for their respective authorities. Unable to reconcile with the perpetrators, the leaders of both communities are also confronted with the dilemma of whether to send the families back to their hometowns, which may be more dangerous, or sustain the families indefinitely as temporaries and as migrants. However, the Sudanese government's recent efforts to initiate a national dialogue, which calls for the formal integration of the displaced families through allocation of residential property, signals a genuine commitment toward more durable solutions, though the families understand this proposal as having diminished their prospects of return. Similar to the proposed solution for the KP families, which calls for securing a dedicated area within the Valley, the proposed national integration policy of the Sudanese government has created some confusion within Darfur families. Darfur family-supported groups view voluntary repatriation as a priority, similar to the advocates of the KP community, who view the "dignified return of the families" as the only solution that honors the wishes of the families. However, given the disconnect between the leaders and the family advocates, the fate of the displaced families remains in limbo for both KPs and the Darfur families.

Given the nature of prolonged displacement, the policymakers of both groups remain challenged in how best to address the family issues, given that the nature of issues has been changing as the family priorities change. However, in both cases, the families are attentive to the fact that the policy portfolio remains a short-term approach, aimed at providing them with temporary relief until the officials reconcile with their hometown leaders and confront the perpetrators.

Contrasts

Although the two cases exhibit similarities that come to be generally associated with long-term displacement, marked differences in the two cases are worth noting as they impact the outcome of IDP policies and the current predicament of the displaced families.

Among the most striking difference in the handling of displacement has been the absence or the presence of external agencies. The crisis in Darfur led to the displacement of more than three million people, calling for a large-scale response, thus necessitating the intervention of international agencies. Consequently, the security of the IDP camps came under the focus of the United Nations, and humanitarian assistance was delivered through international agencies such as the International Committee of Red Cross (ICRC). In the case of the KP displacement, from the outset, the officials had discouraged the involvement of external agencies, labeling the displacement as a sensitive issue of an internal (sovereign) and temporary nature, thus ruling out involvement of external agents or supporters. This striking difference in the two cases explains why displacement of the KP community has remained less understood by the international community, whereas the issue of Darfur had come to occupy a place on the international radar for some time.

However, in spite of international intervention, the Darfur crisis had produced a phenomenon of "homeless or street children" (Huaxia, 2017), resulting in a large pool of children left wandering the streets of Khartoum. Consequently, these children are growing up in unhealthy environment, finding refuge in office basements and sleeping behind shops. The most troubling outcome has been that these children have grown up without schooling. In contrast, all Kashmiri children were provided schooling either on the camp premises or in the host communities, extended under the *Prime Minister's Package of Special Allocation for the Kashmiri Migrant Families.* This initiative not only kept the children away from the streets but also, more importantly, has yielded dividends for both the lawmakers and the families.

The Kashmiri youth now enjoy a comfortable standing in their host communities, contributing as teachers, administrators, and accountants. No children of the Kashmiri families were seen to be begging on the streets. A related struggle for Darfur families has been the remote locations of the camps, which have kept those of working age cut off from the mainstream and locked out of livelihood opportunities. It is feared that Darfur's youth residing in camps may be unable to sustain themselves if the camps were to shut down.

Conclusion

After nearly two decades, both the KP and the Darfur displaced communities remain preoccupied with making meaning of their lives, trying to chart out a future amidst uncertainty, and suffering the consequences of being labeled as displaced and as migrants. In numerous ways, the insights gained through an examination of the Darfur families has enabled a more holistic appreciation of the continued predicament of the KP community and at the same time has illustrated the remarkable self-reliance and resiliency of these families.

Capitalizing on the multipronged research that investigated the displacement of the ethnic KP community, juxtaposed with the challenges of those displaced from Abkhazia and South Ossetia, Darfur, Kosovo, and the Nagorno-Karabakh regions, the next and final chapter is an effort to trace and identify the core IDP experience while recognizing variations in the context-based lived-in experiences of the families. Skillfully extracted from the diagnostic tools used, the chapter showcases the "Best Practices," thus opening a platform for public debate where the role of civil society, intricacies of IDP/Host dynamics, moral hazards of actor positions, expectations of return, and the context-specific role of international actors can be better understood to inform the societal and policy solutions. This effort is crucial for rebuilding communities drenched in political and community entrapment, resulting from their extended internal displacement.

References

Avruch, K. (2011). *Context and pretext in conflict resolution: Culture, identity, power, and practice.* Boulder, CO: Paradigm Publishers.

Barakat, S. (2014, May 29). The potential of local conflict resolution in Darfur. *Christian Science Monitor.* Retrieved from www.csmonitor.com

Brookings-Bern. (2008, April 8). Brooking-Bern Project on Internal Displacement – Annual Report 2007. *Brookings.* Retrieved from www.brookings.edu

Campo, K. (2016, February 25). Darfur is sick of fighting. *Global Post*. Retrieved from www.globalpost.com

Dugan, M. (1996, Summer). A nested theory of conflict. *Women in Leadership, 1*(1), 9–19.

Galtung, J. (1969). Violence, peace, and peace research, *Journal of Peace Research, 6*(3), 167–191.

Harré, R., & van Langenhoven, L. (1999). *Positioning theory: Moral contexts of intentional action*. Malden, MA: Blackwell.

Huaxia. (2017, February 22). Sudanese NGO encourages homeless children to join school. *New China*. Retrieved from http://news.xinhuanet.com

Korostelina, K. (2008). Identity conflicts: Models of dynamics and early warning. In D. Sandole (Ed.), *A handbook of conflict analysis and resolution*. London, UK: Routledge.

Munzoul, A. (2004, July). *Displaced persons in Khartoum: Current realities and post-war scenarios*. A Report for ME Awards, The Population Council, Cairo, Egypt.

Newvision. (2014). Almost 7 million need aid in Sudan. Retrieved from www.newvision.co

Osman, E., & Sahl, I. (2000). *Displacement and poverty: A situation analysis*. A report for Agency for Cooperation and Research in Development (ACORD), Khartoum, Sudan.

Rajput, S. (2012). *The displacement of the Kashmiri Pandits: Dynamics of policies and perspectives of policymakers, host communities and the Internally Displaced Persons*. (Doctoral dissertation). Fairfax, VA: George Mason University. ISBN: 9781267843333.

Rothbart, Daniel. (n.d.) *Internal displacement: Forgotten facet in Darfur conflict*. George Mason University. Unpublished.

Rothbart, D., Brosche, J., & Yousif, A. (2012). Darfur: The crisis continues. *Beyond Intractability*. Retrieved from beyondintractability.org

Salih, K., (2008). The internationalization of the communal conflict in Darfur and its regional and domestic ramifications: 2001–2007. *Arab Studies Quarterly, 30*(3), 1–24.

World Without Genocide. (2013). *Darfur genocide, 2003–present*. Retrieved from http://worldwithoutgenocide.org

Xuequan, Mu. (2015). *Improved security in Sudan's Darfur lures IDPs to return*. Retrieved from news.xinhuanet.com

Section V
Findings, best practices, and moving forward

14 Findings, best practices, and moving forward

Introduction

Capitalizing on the investigative enquiry that unfolded the conflict-induced displacement of the Kashmiri Pandit (KP) community, a high-caste Hindu society (Rajput, 2012), and the subsequent determination to position the KP displacement in the context of similarly displaced communities of Azerbaijan, Georgia, Serbia, and Sudan, this final chapter empowers readers across disciplines with over a dozen findings and best practices and issues a call for action for distinct levels of leadership.

In the context of the KP displacement, the research explored two key perplexing issues of protracted displacements. First, the nature and the impact of elite *positions* on the outcome of IDP policies and, second, the dilemma that keeps the families ambivalent on the issue of return as, despite the promise of robust return and rehabilitation packages, they remain reluctant to return to their hometown communities and yet equally reluctant to embrace their new host communities.

Recognizing the variations that come to shape the families' displacement experience, driven by their context-based experience of residing in the shadows of their host communities, the findings are appropriately labeled "KP specific findings" and "Broad-based findings." Best practices and the lessons learned in the context of the KP displacement and validated by journeying into similarly displaced communities are deemed highly relevant in the context of similarly displaced communities around the globe. The shared challenges of these five distinct communities, as unveiled through the diagnostic tools used to analyze the issues of these communities, namely, Dugan's Nested Model (Dugan, 1996), represent an effort toward identifying the "core displacement experience" of those who must endure extended displacement.

This effort is hoped to elevate the phenomenon of internal displacement beyond the issues and responsibilities of an "internal nature" relegated to a realm of national concern to issues of a universal nature and thus of a global concern. Such an understanding of internal displacement is expected to open possibilities for distinct approaches into context-driven policies and societal solutions.

In numerous ways, not only do the commonalities and contrasts unfolded through examination of these geographically, economically, politically, and culturally dispersed communities yield a deeper appreciation of the predicament of the KP community while pointing to the resiliency and the resourcefulness of this community, but also, more importantly, the comparative discussion has heightened the sensitivity for all those who remain in protracted displacement.

The findings are expected to open a platform for multilevel public debate where the roles of national and international actors, responsibilities of the civil society, and the rightful place for the state and the municipal agencies can be discussed, debated, and better understood. In addition, the findings open possibilities for a dialogue, required to gain a well-versed understanding of the responsibilities and rights of the Host communities; the politics and the intricacies of the IDP/Host dynamics, the moral hazards of actor positions; and the dilemma and the meaning of return. These are central dimensions of protracted displacements and deserve a robust and good-spirited debate in order to guide the context-sensitive policy and community solutions.

IDP phenomenon

As exposed in the opening chapter and worth noting again, the extent of the crisis of those who remain displaced within their national borders, the internally displaced persons (IDPs), is an unsettling phenomenon. As of 2017, about 40 million people across continents remained outside of their hometowns, bearing an average length of 17 years in displacement (IDMC, 2016; OCHA, 2015). Among the forces that result in displacement, conflict-induced displacement is suggested to leave the most enduring scars not only by affecting those who remain in exile but also spilling onto a whole new generation born and raised inside the camps. As unveiled earlier, a systematic grasp of the plight of these communities remains less understood due to the "internal" nature of such crisis, attached to a nation's sovereignty, which assigns the responsibility for their care and protection to national leadership. Relatedly, the national leaders feel obligated to downplay the IDP crisis by either reframing, dismissing, or masking it in order to keep the larger populations undisturbed and the affected populations "protected." However, the examination of the families displaced from the Abkhazia and South Ossetia regions, Darfur, the Kashmir Valley (Valley), Kosovo, and the Nagorno-Karabakh regions makes it clear that in the name of "protection," the reframing of the real crisis of displacement has the effect of keeping the displaced groups "hidden" and the local populations "unconcerned" with the IDP needs, thus relegating the displaced communities to the forgotten and the overlooked communities.

Overview of Kashmiri Pandit displacement

Approximately a quarter of a million people of the minority Hindu KP community, residents of the Kashmir Valley within the Indian state of Jammu and Kashmir, were displaced as a result of the aggressive anti-India campaign in 1989 led by the Muslim majority of the Valley. From the outset, the displacement of this community officially came to be treated under the rubric of "sensitive issues" within the context of the wider "Kashmir problem." Simultaneously, the official dubbing of this community's displacement as a "voluntary migration" kept the crisis from being flagged as an issue of humanitarian and urgent concern. Consequently, after more than 26 years of exile, only random facts and one-sided explanations of the community's ouster have been unveiled and understood. Critical issues had remained hidden or misunderstood, such as the roles of the national leaders and Kashmir's civil society in the interpretation and the management of the crisis, the nature and the rationale of the politics of the IDP/Host dynamics, and the mind-set behind KP policymaking as well as the dilemma of return for both the policymakers and the families. A multipronged investigation of the KP displacement has now unfolded the multiplier effect of the positions and perspectives of the key actors of this displacement (Rajput, 2012). The actor positions, woven around the narratives embraced by each constituency, illustrate how *positions* (Harré & van Langenhove, 1999) come to weigh on the continued predicament of the families, already overwhelmed by the overlapping challenges of an individual, relational, economic, and political dimension. The crossover influence of different actors as well as the multidimensional understanding of the KP displacement gained through this effort is offered here, through the "Findings" presented later.

142 *Findings, best practices, & moving forward*

Key finding

IDP policies are a direct function of elite positions

The probe into the KP displacement and the understanding of displacements of the four distinct communities around the globe makes it clear that the IDP policy formulation becomes and remains a function of how the officials come to address and position the crisis of displacement as it unfolds. This elite *positioning* (Harré & van Langenhove, 1999) is immediately leveraged to assign the labels to be attached to the displaced families, such as the *displaced, migrants, minorities, guests, outsiders,* or *temporaries*. In turn, and going forward, the interpretation of the crisis and these family labels are used to inform the IDP policies. The explanation of displacement as a matter of "temporary disturbance" is correlated to the labeling of the families as *temporaries*, thus justifying policies of a transient and a temporal nature aimed at providing temporary relief (Rajput, 2012). The multiplier effect of the positions takes place when the same elite narratives come to be adopted by members of the host communities. The locals seize on the elite labels to set in motion their own rules, expectations, and boundaries (Tilly, 2005) which are applied to the new arrivals into their communities, which will be used to govern every aspect of the life of the newcomers. This insight provides a unique lens into the political (through positioning) and the community entrapment (through the social ordering of the community) of those who remain displaced within their national borders.

In order to facilitate a grasp of the KP displacement in particular and of the global phenomenon of protracted displacement in general, the subsidiary findings are unpacked later as "KP specific findings" and "Broad-based findings."

KP specific findings

Positions and labels impede long-term solutions

The official dubbing of the crisis that led to the displacement of the KP families as a "temporary disturbance that triggered a voluntary migration" (personal communication, August 1, 2011, Delhi-based government official) has helped to justify a temporary fix, thus ruling out any need for long-term solutions. This was reflected through the government's effort to retain the ownership of the "migrant shops and the townships" (Rajput, 2012) made available for the families on a temporary basis. The "temporary disturbance" explanation also allowed the lawmakers to experiment with ad hoc policies that came and disappeared, while the families remained in exile. Additionally, positioning of this displacement as a "voluntary migration" liberated the leaders from confronting the perpetrators and demanding justice for the families' ouster and for exploring policies for the families' return. Not only did the "migrant" label dictate how the KP families were to be treated by the top tier, but also the same narrative transferred to the psyche of the members of the Host communities and quickly introduced hierarchy, establishing "rules for inclusion and exclusion" (Tilly, 2005), restricting "outsiders" from access to community resources.

Conflicting elite positions backfire and empower the IDPs

Maintaining two contradictory positions simultaneously, namely: that the families are to return "when normalcy happens" (Rajput, 2012); while claiming that the families will not return, given the benefits of the metro societies has backfired. Although the conflicting narratives are likely meant to favor and protect the elite, paradoxically, the same narratives have come to empower the IDP psyche. Capitalizing on the mixed signals, the KP families use their agency to position themselves to demand full rehabilitation in new communities, sensing that the government doubts that they will go back; based on the dominant narrative that "the families are to go back," the families press for their return, demanding incentives for doing so, thus skillfully keeping the elite on the defensive.

Civil society becomes scapegoats for flawed policies

Immediately after the KP displacement, with the proliferation of the KP-based nongovernmental organizations (NGOs), numerous platforms became available for the families to express their voices. Initially, some organizations came onto the scene to assist the families with administrative tasks, such as the registration process, or to help them navigate their new society. However, fatigued by the lengthy exile and the KP families' evolving needs, the work of these groups often took on ambiguous and contradictory positions. At times, the advocacy groups argued for the safe return of the families to their Valley ancestral hometowns, while other family-supported groups argued for the return of the families to a newly constructed separate homeland away from the Valley. Capitalizing on the mixed signals from the family supporters, the officials found it easier to fix the blame for any flawed or premature policies on the misrepresentation of the families' needs by their own groups.

Protective IDP policies stifle IDP/Host dynamic

The government-arranged, "Migrant Township"-like settlements provided to the KP families in Jammu or the Tserovani camps provided for those displaced from the Abkhazia and the South Ossetia regions of Georgia come to be justified as arrangements that uniquely provide a "safe and home-like experience" for the families. Such measures are said to facilitate the easier return of the families. Segregated housing is promoted as a means of preserving the family culture, thus minimizing the hardships of adapting to the larger communities. Not only do these overprotective policies come to deprive the families of economic opportunities, but such strategy also stifles the community's social integration. These physical boundaries come to inadvertently reinforce the stereotypical image that each group holds of the other, thus giving rise to the advent of parallel communities.

Useful policies yield strong IDP/Host dynamic

There is a positive correlation between the favorable assessment of policies by the families and the resulting IDP/Host dynamic. The KP families whose pre-displacement government salaries were protected after their displacement, found it feasible and even gratifying to share their skills, knowledge, and their free time to strengthen and promote the work of their host communities in Delhi. This outreach, on the part of the displaced families, came from an acknowledgement that these new communities may be their permanent residence. The contributions from families came in the form of providing tutoring services in the schools of their children, helping at the local library, teaching the local children about Kashmir's culture, and writing newspaper articles for community papers (Rajput, 2012). Thoughtful engagement within the host communities allowed the KP families to validate their own sense of worth and identity (Korostelina, 2007). These initiatives not only helped the families secure their own place in the new societies but, more notably, also motivated the host communities to adjust and relax their own "social boundaries" (Tilly, 2005), thus offering the KP families inclusion and membership in their communities.

Subcultures and sub-identities dominate in displacement

Historically, the KP families and the host communities shared the same national and religious identities and ways of life under the overarching umbrella of being Indian citizens of Hindu faith. However, once displaced, the KP families organized their lives around the newly assigned "migrant" identity, downplaying the broader national identity. The families began to define themselves more in terms of KP traits rather than as Indian Hindus. This was partly a consequence of the government-arranged housing that led to the perception that perhaps the IDP and the Host communities shared little in common. However, it also reflected the choices made by the KP communities, which began to project their own group as distinct and dissimilar to the host group (Tajfel & Turner, 1979). This often led the KP families to mock the lifestyle of the locals of Delhi and Jammu as diluted and immoral, pointing to their practice of mixed marriages and nuclear families as values inferior to those practiced by their own KP community. Embracing the subcultures and sub-identities in place of the larger identity led to a limiting of the families' interactions with the wider community, justified by the KP families as a mechanism for keeping them protected from the unnecessary pressures of the host community and as a device to uphold their superior values.

Broad-based findings

The exclusive insights, gained through the probe of KP displacement and validated by journeying into the communities of Azerbaijan, Georgia, Serbia, and Sudan (Chapters 10–13), yield a holistic understanding of the global phenomenon of internal displacement. These insights facilitate the tracing of the commonalities and contrasts between distinct communities, thus elevating the grasp of displacement in terms of core displacement experience. The following broad-based findings are aimed to open opportunities for possible approaches into societal and policy solutions.

Displacement must be understood as a multidimensional phenomenon

Internal displacement must be understood as a complex conflict issue as it touches and adversely magnifies all aspects of those displaced, such as the individual, social, economic, legal, and political (Rajput, 2015). The phenomenon must be understood in terms of the Nested Model (Dugan, 1996) as a comprehensive system comprising of interconnected and entrenched elements (Figure 5.1). Specifically, the causes of conflict-induced displacement must be recognized as stemming from a breakdown in a society's structure (societal dimension), which result in the loss of one's dwelling (physical dimension).

Subsequently, the loss of home requires complex transitioning into a new society (relational, social, economic, and psychological dimensions of displacement). The rules of the new community are embedded in the society's overarching systems that make up the rules and procedures that govern the boundaries and the rights for the IDPs (political and legal dimensions). Such a holistic understanding underscores the need to address the issues of internal displacement with a similar holistic mind-set and policy perspective. Failure to appreciate the all-inclusive nature of the challenges of those displaced is dangerous as it leads to ad hoc policies, some with unintended consequences and moral hazard, and some yielding only minimal impact.

Succeeding generations share displacement burden

Mass exodus of people forcefully evicted from homes breaks up families; cuts social and cultural connections; and jeopardizes employment, education, and marriage opportunities for many. Just as the mass exodus of KP families had inflicted severe hardships not only on those directly affected but also on the succeeding generation that grew up in the "township" communities, the youth of those evicted from Abkhazia and South Ossetia, Kosovo, Nagorno-Karabakh, and Darfur also remain tied to the hardships of the process of rebuilding their lives. In long-term displacements, the succeeding generations find themselves at a loss, unable to fully grasp the pain of the atrocities and humiliation inflicted on their elders.

Resistance to return transcends the generation gap

The idea that the youth brought up outside the ancestral communities and shaped by a culture of metropolitan communities stands in the way of the return of their elders takes a different meaning in cases of conflict-induced displacements. Undeniably, there are Abkhazians, ethnic Serbs, and KP elders who yearn to return to their hometowns; however, in these communities there are also elders who are "repelled" by the very idea of going back to the communities that "humiliated their identity" (Rajput, 2012). These families express disgust at the idea of returning to "live amongst those who ousted them from their communities" (personal communication, July 27, 2011, KP family at Dwarka camp). The youth, being shaped by the perspective of their elders, who convey a humiliating experience of their eviction, become equally repelled by the idea of visiting a place that humiliated their grandparents. Therefore, specifically in the event of conflict-induced displacements, more so than displacements from natural disasters, resistance to return becomes a personal issue for the youth rather than a simple matter of the generation gap, or a simple cost-benefit consideration, given the opportunities of a metropolitan society.

Security and trust remain the key concerns for returnees

Although, while in displacement, the families admit to not being "able to live the lives that they had planned and dreamed of" (personal communication, May 13, 2015, Abkhazian IDP, Tserovani camp) and "miss the brotherly love of the Valley neighbors" (personal communication, July 18, 2011, KP, Jagti camp), the fear of return remains the most enduring of the displacement issues. As the years in displacement add up, the number of families willing to return begins to decline. Even those with their "hearts in Kashmir" admit that the social fabric of their society has changed forever, the "seeds of mistrust have been sown," and "society can never be trusted again" (Rajput, 2012), and other displaced communities have similar concerns when contemplating the idea of return. Ethnic Serbs fear the continued marginalization of their group (personal communication, May 6, 2016, displaced Serbian at Belgrade Collective Center), and the Abkhazian families fear being "targeted by the secessionists." The families remain apprehensive of the security in their hometowns as the presence of security forces creates the perception that perhaps their hometowns are not yet peaceful and ready for their return.

Overall, the displaced communities of Azerbaijan, Georgia, India, Serbia, and Sudan exhibit a peculiar commonality, reflecting a state of ambivalence and the continued unrest while remaining in extended internal exile. On the one hand, having been displaced, these families speak of a complete loss of identity and homelessness, yet simultaneously these families cling to the same identity that keeps them hopeful of returning to the very communities that humiliated and ousted them and their identities.

Best practices and way forward

A consideration of the best practices observed through the journeying into the five distinct communities, representing the larger IDP phenomenon, is expected to open a platform for multilevel public debate. Such a platform will allow discussions on issues of critical importance for families in extended displacements, such as the ethics of labeling those displaced, the politics and intricacies of IDP/Host dynamics, the

148 *Findings, best practices, & moving forward*

roles of civil society, and the partnering of national and international actors. These aspects of the "displacement system" can fruitfully guide the context-sensitive societal, political, and legal reforms, capable of installing meaningful and durable solutions for communities that must endure displacement for indefinite periods. The following section details the issues that deserve to become the focus of thoughtful public discussion and dialogue.

Changing the course on IDP labels

As understood from the KP displacement and from the labeling of those displaced from the Nagorno-Karabakh region, the labels and the categories assigned to those displaced becomes a direct function of how the crisis of internal displacement is interpreted and addressed at the top level. The labels, so assigned, trigger a "multiplier effect" (Rajput, 2012) in not only subscribing a new identity to the families but also sending signals to members of the host communities, who immediately begin to strategize on how the displaced are to be treated. The labeling of the displaced families as "temporaries," "migrants," or "outsiders" sends prompt signals to members of the host community, school personnel, and law enforcement agencies as to how the new arrivals are to be treated and, specifically, which services and resources are to be made available and which are to be restricted. The markers, so assigned, become the lens through which all actions of the families come to be assessed. These labels help the host communities to perceive the displaced as somehow less than whole, as "freeloaders" and "lazy," instead of receiving them as those who bring extra resources to their communities and as their co-citizens. After multiple decades, as the elite remain committed to the labels and their "positions" (Harré & van Langenhove, 1999), the families remain stigmatized and humiliated by the same. A thoughtful examination of the labeling and the categorization, specifically of those who remain in protracted displacement, needs to take place at the national, international, and local levels, and incorporated within the framework of the *Guiding Principles on Internal Displacement*. This suggests a context-appropriate engagement and guidance of international actors.

Fostering IDP/Host dynamics

In the context of long-term displacements, where the hope of return starts to wane as years and decades add up, it is crucial that both the IDP and the Host communities have "access to opportunities that enable them to connect and build mutual bonds" in the interest of promoting the goals of the wider community (Allport, 1954). Keeping the displaced communities isolated in township-like settlements, decade after decade, rules out possibilities for such openings. As demonstrated by the engagement of the KP families who have helped advance the work of their larger Delhi community, through which they have demonstrated their own resilience, such initiatives can boost the self-worth of the families and simultaneously give the host communities convincing reasons to adjust their own perceptions. To foster such IDP/Host dynamic for certain communities, the engagement of civil society may be necessary.

Educating the youth

One of the key praiseworthy outcomes of the policies formulated for the KP families has been the access to education by the KP children through the initiatives supported by the Delhi-based central government. Over and above any other policies designed for the KP families, the families place the benefits gained from the centrally supported education at the highest level (Rajput, 2012, 2016). Owing to the education policy, some families have come to frame their displacement from the horrific ouster of their community as a "blessing in disguise." Not only did education for the KP youth keep the young minds occupied and away from the streets, but, more importantly, it has empowered the families with the survival tools needed to navigate and cope with the trauma of displacement as well as to meet the demands of their new society. This contrasts with cases of displacements elsewhere, where absence of adequate access to education for the children of the IDPs has led to child labor and exposure to vulnerabilities (Yilmaz, 2003). Adoption of comparable education policies for displaced communities elsewhere deserves thoughtful consideration.

Partnering of national and international actors

The scale of internal displacement and the average length of family exiles underscore the need as well as a call for action to yield a stronger national-international partnership. In the absence of a dedicated agency, the care and protection of the internally displaced families is left to the whims and priorities of the national actors, who often struggle with the interpretation, acknowledgment, and handling of the crisis, fearing the outcry from the larger population. This often results in the dubbing of the crisis in secretive ways so as to disguise the societal problems that occur under their watch, thus projecting the façade of a peaceful society. A misrepresentation of the crisis results in the crafting of ill-suited and arbitrary policies. As exile protracts, and the families remain hidden, any urgency to address and resolve the crisis gets pushed further from the policy arena. At the same time, the families remain unsure of whether they should begin to embrace their new society or begin to prepare for return. The fact that they are not able to return to their homes after more than two decades, despite the seemingly attractive return and rehabilitation packages offered by national governments, suggests a guiding and an advising role for the international actors. Such guidance can come in the form of assisting the national leaders in recognizing and acknowledging the crisis of internal displacement as it is, as opposed to misdiagnosing the crisis, leading to ill-suited and ineffective prescriptions, thus keeping the lives of upwards of 40 million people in limbo, eroding of their human worth.

Moving forward

Given the protracted nature of conflict-induced displacements, over time, the individual issues come to be meshed and embedded into other issues, thus obscuring a clear understanding of the families' continued predicament. The schema of IDP issues uniquely framed around the Nested Model (Dugan, 1996), showcased throughout this book, should form an integral component of the diagnostic tool kit to analyze displacement issues across communities. Failure to appreciate the holistic and embedded nature of the challenges of those who remain in extended exile is dangerous as it leads to conflicting policies, often yielding unintended consequences and moral hazard for both the lawmakers and the families.

Core IDP experience can now be understood to be woven around political and community entrapment, and should be explained as follows:

> Evicted from homes, and having become homeless, the families are tasked with entering a tricky and often politically dictated transition and transformation into host communities. Next, the government-arranged accommodations (political entrapment) and members of the Host community (community entrapment) work in parallel to shape the IDP experience by regulating the IDP/Host relations. While subjected to the social order of the new society, often down-playing own values and beliefs, the families remain uncertain of the institutional systems that may or may not guarantee their return or secure their integration into the new communities, thus remaining assigned to a void for indefinite periods (dilemma of return).
>
> (Rajput, 2012)

References

Allport, G. (1954). *The nature of prejudice.* Cambridge, MA: Perseus Books.

Dugan, M. (1996, Summer). A nested theory of conflict. *Women in Leadership, 1*(1), 9–20.

Harré, R., & van Langenhove, L. (1999). *Positioning theory: Moral contexts of intentional action.* Malden, MA: Blackwell Publishers.

IDMC. (2016). *Grid 2016. Global report on internal displacement.* Internal Displacement Monitoring Center, Norwegian Refugee Council, Oslo, Norway.

Korostelina, K. (2007). *Social identity and conflict: Structures, dynamics, and implications.* New York, NY: Palgrave Macmillan.

OCHA. (2015, January 22). *The forgotten millions.* Geneva, Switzerland: United Nations Office for the Coordination of Human Affairs.

Rajput, S. (2012). *The displacement of the Kashmiri Pandits: Dynamics of policies and perspectives of policymakers, host communities and the Internally Displaced Persons* (Doctoral dissertation). Fairfax, VA: George Mason University. ISBN: 9781267843333.

Rajput, S. (2015). Chapter 3: Internal displacement of Kashmiri Pandits. In S. Kukreja (Ed.), *State, society, and minorities in South and Southeast Asia* (p. 62), New York, NY: Lexington Books.

Rajput, S. (2016). Transitional policies and durable solutions for displaced Kashmiri Pandits, *Forced Migration Review Issue 52.* Refugee Studies Center, Oxford University, London, UK.

Tajfel, H., & Turner, J. (1979). An integrative theory of intergroup conflict. In W. G. Austin & S. Worchel (Eds.), *The social psychology of intergroup relations* (pp. 56–61). Monterey Bay, CA: Scientific Research.

Tilly, C. (2005). *Identities, boundaries and social ties.* Boulder, CO: Paradigm Publishers.

Yilmaz, B. (2003, February 23). Forced migrants in Istanbul. *Researching internal displacement: State of the art. Conference report* (p. 36), Norwegian University of Science and Technology, Trondheim, Norway.

Annex 1
Fieldwork log

The displacement of the Kashmiri Pandits: dynamics of
policies and perspectives of policymakers, host communities
and the Internally Displaced Persons
Doctoral dissertation, George Mason University, USA
Fieldwork Log
July–August 2011
S. Rajput

Interview #	Interview Date	Location	Actor	Affiliation
1	July 18	Jammu	IDP	Jagti Camp Resident
2	July 18	Jammu	Pol Imple.	Pd – All Migrant Camp Coordination Comm.
3	July 18	Jammu	Pol Imple.	Commissioner for Relief & Rehab
4	July 18	Jammu	Host	Taxi Driver
5	July 18	Jammu	Host	Bureau Chief, "Times Now"
6	July 19	Jammu	IDP	Muthi Camp
7	July 19	Jammu	IDP	Vishasta Charity Hospital
8	July 19	Jammu	NGO	Chairman, Bhartiya Seva Samiti
9	July 20	Jammu	Host	Dogra family
10	July 20	Jammu	IDP	Jagti Camp
11	July 21	Srinagar	Policy Maker	Member of Legislative Assembly
12	July 21	Srinagar	IDP	Veesu Camp – Returning Teacher
13	July 21	Srinagar	Host	Muslim Family
14	July 21	Srinagar	Pol Imple.	Former District Commissioner
15	July 22	Srinagar	Po Imple.	Chief Minister's Office – Office of Revenue

(*Continued*)

Interview #	Interview Date	Location	Actor	Affiliation
16	July 23	Jammu	Host	Dogra Businessman
17	July 23	Jammu	IDP	Jagti Camp
18	July 23	Jammu	IDP	Jagti Camp – Shopkeeper
19	July 23	Jammu	IDP	Muthi Camp – Musician from Valley
20	July 24	Delhi	IDP	Random Enquiry
21	July 24	Delhi	NGO	Random Enquiry
22	July 25	Delhi	NGO	President, Kashmir Samiti
23	July 25	Delhi/Dwarka	Host	Media in Dwarka Camps
24	July 26	Delhi/INA	IDP	Shopkeeper 1 – INA Migrant Market
25	July 26	Delhi/INA	IDP	Shopkeeper 2 – INA Migrant Market
26	July 26	Delhi/INA	IDP	Shopkeeper 3 – INA Migrant Market
27	July 26	Delhi/INA	IDP	Shopkeeper 4 – INA Migrant Market
28	July 26	Delhi/INA	IDP	Shopkeeper 5 – INA Migrant Market
29	July 26	Delhi/INA	Host	Local customer at INA Shops
30	July 27	Delhi/Dwarka	IDP	Dwarka Camps Sector 1
31	July 27	Delhi/Dwarka	IDP	Dwarka Camps – President, Welfare Assoc.
32	July 27	Delhi/Dwarka	IDP	Dwarka Camps
33	July 28	Delhi	NGO	President, Kashmir Samiti
34	July 28	Delhi	IDP	Kashmiri Samiti
35	July 28	Delhi	NGO	Chief Spokesperson, Pannun Kashmir
36	July 29	Delhi/Rohini	IDP	Family 1
37	July 29	Delhi/Rohini	IDP	Family 2
38	July 29	Delhi/Rohini	IDP	Shopkeeper
39	July 30	Delhi/Rohini	IDP	Family 3
40	July 30	Delhi/Rohini	Host	Host Family 1
41	July 30	Delhi/Rohini	Host	Host Family 2
42	July 30	Delhi/Rohini	Host	President, Welfare Association
43	July 30	Delhi/Rohini	Host	Office Worker
44	August 1	Delhi	Policy Maker	Joint Secretary, Ministry of Home Affairs

Interview #	Interview Date	Location	Actor	Affiliation
45	August 1	Delhi	Policy Maker	Media Director, Ministry of Home Affairs
46	August 2	Delhi/Bappu Dham	IDP	Family (Previous Camper)
47	August 2	Delhi/Bappu Dham	NGO	Secretary, Delhi Pradesh Youth Congress
48	August 2	Delhi/Bappu Dham	Host	Host
49	August 2	Delhi/Bappu Dham	Host	Community Activist
50	August 2	Delhi/Bappu Dham	IDP	Random Interview on Street
51	August 3	Delhi	Pol Imple.	Principal, Kendriya Vidalaya School
52	August 3	Delhi/Pomposh Enclave	IDP	Family, Professor from Valley
53	August 4	Delhi/INA	IDP	INA Migrant Market
54	August 4	Delhi/INA	Local	INA Migrant Market (Local Customer 1)
55	August 4	Delhi/INA	Local	INA Migrant Market (Local Customer 2)
56	August 5	Delhi/Krishna Mkt	IDP	Family 1
57	August 5	Delhi/Krishna Mkt	IDP	Family 2
58	August 5	Delhi/Krishna Mkt	IDP	Family 3
59	August 5	Delhi/Krishna Mkt	IDP	Family 4
60	August 6	Delhi/Pomposh Enclave	NGO	Chairman, S. K. Foundation
61	August 6	Delhi/Pomposh Enclave	IDP	Family 1
62	August 6	Delhi/Pomposh Enclave	IDP	Librarian, Kashmir Studies Center
63	August 8	Delhi	Pol Imple.	Commissioner, Delhi Development Authority
64	August 8	Delhi/INA Mkt	IDP	Customer at Store
65	August 8	Delhi/INA Mkt	IDP	INA Migrant Mkt.

(Continued)

Interview #	Interview Date	Location	Actor	Affiliation
66	August 9	Delhi	Policy Maker	Minister of J&K Affairs, Former Chief Minister (Abdullah)
67	August 9	Delhi/Yusuf Sarai	IDP	Shopkeeper 1 – Yusuf Sarai Kashmiri Mkt
68	August 9	Delhi/Yusuf Sarai	IDP	Shopkeeper 2 – Yusuf Sarai Kashmiri Mkt
69	August 9	Delhi/Yusuf Sarai	IDP	Shopkeeper 3 – Yusuf Sarai Kashmiri Mkt
70	August 9	Delhi/Yusuf Sarai	IDP	Doctor – Yusuf Sarai Kashmiri Mkt
71	August 9	Delhi/Yusuf Sarai	IDP	Shopkeeper 4 – Yusuf Sarai Kashmiri Mkt
72	August 11	Delhi	Policy Maker	Chairman, National Commission on Minorities
73	August 12	Delhi	Pol Imple.	Under Secretary, Kashmir State House
74	August 12	Delhi	Pol Imple.	Assist Director, Migrant Cell, Kashmir House
75	August 13	Delhi	Pol Imple.	Director, Planning, Kashmir House
76	August 13	Delhi	IDP	Employee 1 at Kashmir House
77	August 14	Delhi	IDP	Employee 2 at Kashmir House
78	August 15	Delhi/Dwarka	IDP	Dwarka Camps, Sector 5
79	August 16	Delhi/Dwarka	IDP	Family 1 at Ramphal Community
80	August 16	Delhi/Dwarka	IDP	Family 2 at Ramphal Community
81	August 17	Delhi/Pitam Pura	IDP	Secretary of Association
82	August 17	Delhi/Pitam Pura	IDP	Family 1
83	August 18	Delhi/Ali Ganj Camp	IDP	Shopkeeper from Migrant Mkt
84	August 18	Delhi/Ali Ganj Camp	IDP	Proxy for Brother
85	August 18	Delhi/Ali Ganj Camp	IDP	Doctor at the Temple
86	August 19	Delhi/Hauz Rani Shops	IDP	Family 1
87	August 19	Delhi/Hauz Rani Shops	IDP	Family 2
88	August 19	Delhi/Hauz Rani Shops	IDP	Family 3

Interview #	Interview Date	Location	Actor	Affiliation
89	August 19	Delhi/Hauz Rani Shops	Host	Shopkeeper
90	August 22	Delhi	Pol Imple.	Superintendent, Office of Magistrate
91	August 22	Delhi	Pol Imple.	Divisional Commissioner
92	August 22	Delhi	Policy Maker	General Secretary, AIKS
94	August 22	Delhi	IDP	Person 1 at AIKS
95	August 22	Delhi	IDP	Person 2 at AIKS

Annex 2
Kashmiri Pandit families assess 'migrant' policies

The displacement of the Kashmiri Pandits: Dynamics of policies and perspectives of policymakers, host communities and the Internally Displaced Persons
Doctoral dissertation, George Mason University, USA
July – August 2011
S. Rajput

Annex 2
Policy Assessment and Reasons for KP Displacement
Source: Face to Face Interviews – July–August 2011

	Beneficial		Somewhat Beneficial		Failed		Politically Motivated		Unhelpful	
	Delhi	Jammu	Delhi	Jammu	Delhi	Jammu	Delhi	Jammu	Delhi	Jammu
Reasons for displacement										
Lack of security	40%	50%	75%	100%	60%	100%	75%	100%	66%	100%
Humiliation and forced eviction	60%	0%	25%	0%	40%	0%	25%	0%	33%	0%
Dyeing community	0%	50%	0%	0%	0%	0%	0%	0%	0%	0%
Challenges encountered										
Psychological	60%	50%	63%	50%	33%	100%	50%	100%	17%	100%
Enviornmental	20%	50%	37%	50%	40%	0%	50%	0%	50%	100%
Institutional	20%	0%	0%	0%	33%	0%	0%	0%	34%	0%
Intent to return										
No intent to return	60%	50%	38%	100%	67%	100%	40%	100%	50%	100%
Maybe return	40%	50%	63%	0%	33%	0%	60%	0%	50%	0%
Interactions with wider										
Good interactions	80%	0%	25%	0%	33%	0%	40%	0%	33%	0%

Annex 2
Policy Assessment and Reasons for KP Displacement
Source: Face to Face Interviews – July–August 2011

	Beneficial		Somewhat Beneficial		Failed		Politically Motivated		Unhelpful	
	Delhi	Jammu	Delhi	Jammu	Delhi	Jammu	Delhi	Jammu	Delhi	Jammu
Advocacy										
KP-based NGOs	80%	0%	63%	50%	33%	0%	60%	100%	50%	0%
Central government	0%	50%	0%	50%	7%	0%	0%	0%	0%	0%
Other groups	0%	50%	38%	0%	0%	0%	10%	0%	0%	0%
No advocacy or advocacy without results	20%	0%	0%	0%	60%	0%	30%	0%	50%	100%
Changes over the years										
Positive changes	60%	0%	14%	0%	27%	0%	20%	0%	17%	0%
Negative changes	40%	100%	86%	100%	73%	100%	60%	100%	83%	100%

Rajput (2012).

Index

Abkhazia and South Ossetia 89, 141, 144, 146–7; displaced from 101–10
acculturation 31
actors of displacement 9; international 8, 11–2, 15, 17, 27, 82, 109, 120, 134, 148–50; absence of 31–2; national 3, 10, 13, 31, 33, 80, 104
adjustment 34, 104, 108, 113–5, 118–9, 125; adjustment challenge 14, 33, 51, 54, 69, 90, 93–4, 98, 125; Kashmiri challenge 37–40
advocacy 8, 54, 79–8, 87, 143, *see also* Annex 2
All India Kashmiri Samaj (AIKS) 54, *see also* Annex 1, 2
Allport, G. 49, 149
Armenia 91, 96
assimilation 31, 114
autonomous 21, 101, 113, 117, 123, 131
Avruch, K. 8, 38, 47, 57, 60, 118, 125, 131
Azerbaijan 5, 18, 28, 89; displaced from Nagorno-Karabakh 91–9

Belgrade 9, 112, 114, 117–8, 147
British Empire 21, 130
Brosche, J. 123
Brookings-Bern 131

categories 3, 27, 123, 148; categorized 103; categorization 148
civil society 8–9, 15–6, 76, 105, 134, 140–1, 148–9; KP displacement 79–80, 87–8; Darfur 130; Georgia 109; scapegoats 143

collective centers 94–5, 105, 114, 116, 120
collective memory 47, 58
community building 54
compartmentalized communities 4, 28, 47, 104
conflict system 10, 46–8, 51–2, 92, 102, 113–4, 126; complex conflict issues 11, 126
Convention Relating to the Status of Refugees 82
coping skills 37, 98, 126

Darfur 5, 18, 31; displaced from 123–35
Delhi 24, 27–8, 49–51, 61–2, 73, 86, 119–120, 131–2, 144–5, 149; Delhi Development Authority 71
demographic shift of Valley 60
dependence 75, 102, 108; dependency 16, 69
displacement experience 7, 9, 15–6, 64, 69, 73–4, 139, 145; core IDP experience 150
Dugan's Nested Model 14–5, 139, 150; displacement Azerbaijan 91–9; Georgia 102–4; Kashmir 46–56; Serbia 113–5; Sudan 124–7
durable solutions 11, 20, 53, 76, 85, 97, 105, 109, 115, 117, 120, 132, 148
Dwarka camps; *see* Annex 1, 2

ecological settings 34–5
elite positions *see* positions

Index

ethnic: community 88, 101, 118; minorities 22, 35, 97; Serbs 5, 18, 112–21, 147; violence 4
Evans, A. 12, 60
eviction 4, 47–9, 61, 91–2, 98, 146; circumstances of KP 36; forced 7, 21, 53 *see also* Annex 2
exile: internal 4, 97, 101, 108, 112, 147
exodus 21, 23–4, 29, 30, 60, 78, 84–5, 95, 97, 112, 146

Ferris, E. 49
Festinger, L. 7, 59

Galtung, J. 23, 32, 52, 78, 94, 127
generation gap 63, 128, 146–7
Georgia 5, 18, 89, 124, 144, 147; displaced from Abkhazia and South Ossetia 101–10
Guiding Principles on Internal Displacement 3, 10, 83, 148
Gurr, T. 38, 93

Harre, R & van Langenhoven, L. 5, 16, 53, 62, 76, 128, 141–2, 148
hidden communities 12, 24; populations 26
Hindu 4, 22–4, 35, 37, 139, 141–5
home communities 3–5, 7, 14, 17, 45, 77, 85, 87, 117, 119, 132
homeless, homelessness 47–8, 92, 97, 102, 108, 118, 128, 133, 147, 150
host community 15, 27, 38, 150; perspective 62; resistance 85
Host/IDP dynamic 49, 98; *see also* IDP/Host dynamics
IDP/Host dynamic 6, 8, 11, 48, 51, 114, 140–1, 144, 147–9; *see* also Host/IDP dynamic
humanitarian 3, 8, 11, 74, 83, 141; agencies 30, 127; assistance 109, 133; emergency 21

identity 39–40, 47, 58–60, 62, 92, 99, 114, 119, 131, 144–8; "migrant" 45
IDMC 3, 5, 10, 28, 91–3, 95, 99, 102, 104–7, 109, 112–3, 116–7, 140
IDP 3, 13, 19, 83, 140, 143, 146, 149; current debate 10; phenomenon 3, 6, 10, 140, 147

IDP policies: KP 69–75; Azerbaijan 94–95; Georgia 105–116; Serbia 115–116; Darfur 127–128; solutions 139–150
IDP policymaking 6–7, 15, 16, 81, 104, 114; complexity of KP 81–8
India 21–2, 30, 131; anti-India campaign 22, 141; anti-India sentiment 14; anti-India movement 23
institutional violence *see structural violence*
integration 11, 63, 92–3, 99, 104–7, 108–9, 117, 127, 130, 132, 144, 150
international: community 119, 126, 133; absence of framework 83–8; guidance 88
Islam 22–3; Islamic 22–3, 60

Jagti camp 34; Migrant Township 38, 119
joint family system 49

Kashmir Studies Center, Kashmiri Samiti, Kashmir House *see* Annex 2
Khartoum 9, 125, 127–8, 130, 132–3
Korostelina, K. 28, 34, 47, 50, 58, 61, 72, 94, 102, 114, 125–6, 144
Kosovo 4, 9, 18, 90, 110, 112–121

labels 131, 142, 148; labeling 4, 6, 8, 78, 94, 97, 110, 133; misfortunes of 53
legacies KP 34–41
livelihood 52, 72–3, 86, 95, 105–6, 109–10, 116, 123, 128, 130, 134

migrant: community 36, 59; identity 10, 45, 145; market 9, 72; policies 69–75; Township 24, 36, 97, 108, 119–20, 144
migration: voluntary 29, 78, 141–2
militancy 11, 14, 23–4, 37, 52, 131
Ministry of Home Affairs *see* Annex 1
minorities 13, 22–3, 35, 60, 97, 142; community 13, 17, 19, 21–23, 52, 91, 115, 117–9, 123, 131; identities 22
Muslim 22–3, 35, 37, 60, 123, 132, 141

Nagorno-Karabakh: *see* Azerbaijan
narratives *see* positions
national: agenda 17, 81; constraints KP policymaking 84; dialogue 129, 132; discourse 120; identity 145; national-international partnership 149
nationalism 113, 118
NATO 113
Nested Model See Dugan

Oberoi, S. 12, 59–60
Occupied Territories 9, 101, 105

Pakistan 22, 30
Pandita, R. 49–50, 58
Pannun Kashmir *see* Annex 1
parallel societies 51
policies: see IDP policies; *also* IDP policymaking
positions 5–8, 11, 13–14, 39, 54, 57–60, 63–4; civil society 79; elite 29, 76, 142–3; family 79–80; policy outcome 76–81
power relations 6, 53
protracted displacement 5, 8, 10–11, 13–4, 18–9, 24; challenge of researching 26–32

Rajput, S. 7–10, 23–4, 27–30, 49–50, 60–1, 86, 124, 131, 139, 141–50
refugees 3–4, 6, 83, 112
rehabilitation 4–5, 8, 58, 63, 85, 87, 139, 143, 150
relative deprivation *see* Gurr, T.
repatriation 130, 132
return: 4–6, 10, 146–7; Kashmiri dilemma 57–64; Azerbaijan 96–7; Georgia 106–7, Serbia 117–8; Darfur 128–30

Rothbart, D. 124–6, 129
Russian Federation 102

Sazawal, V. 80
security 23, 47, 60, 80, 106–7, 125, 127, 132, 146–7
segregation 77, 93, 104, 110; segregated communities 77
Sekhawat, S. 12, 58
self-reliance 11, 77, 94, 98, 116, 134
Serbia: displacement of ethnic 112–21
South Ossetia *see* Abkhazia *see also* Georgia
sovereignty 3, 13, 140
stigma: 3, 48, 92, 96, 103, 107; stigmatized 28, 39, 77, 97, 101, 120, 148; stigmatizing 74
Structural Violence 52
Sudan: displaced from Darfur 123–39

Tajfel, H. and Turner, J. 145
Tilly, C. 5, 53, 142–4
transformation 33, 150
trauma: chosen 47; collective 103, 108

UNAMID 124, 126
UNHCR 3, 10, 83, 91–4, 96, 98, 104, 109, 116

Volkan, V. 47, 58, 103, 108

Wolpert, S. 22
World Bank 95–6

Yilmaz, B. 149
Yousif, A. 124
Yugoslav Federation 112

9781032929569